Free-Space Optics

Free-Space Optics

Propagation and Communication

Olivier Bouchet
Hervé Sizun
Christian Boisrobert
Frédérique de Fornel
Pierre-Noël Favennec

Series Editor
Pierre-Noël Favennec

First published in France by Hermes Science/Lavoisier entitled "Optique sans fil: propagation et communication"
First published in Great Britain and the United States in 2006 by ISTE Ltd

ISTE Ltd
6 Fitzroy Square
London W1T 5DX
UK
www.iste.co.uk

ISTE USA
4308 Patrice Road
Newport Beach, CA 92663
USA

© GET and Lavoisier

© ISTE Ltd.

First South Asian Edition 2007

The rights of Olivier Boucher, Hervé Sizun, Christian Boisrobert, Frédérique de Fornel, Pierre-Noël Favennec to be identified as the authors of this work has been asserted by them in accordance with the Copyright, Designs and Patents Act 1988.

Library of Congress Cataloging-in-Publication Data

Optique sans fil, English:
Free-space optics : propagation and communication / Olivier Bouchet [... et al.].
   p. cm.
   First published in France in 2004 by Hermes Science/Lavoisier entitled "Optique sans fil: Propagation et Communication."
Includes bibliographical references and index.
ISBN 13: 978-1-905209-02-6
1. Free-space optical interconnects. 2. Optical communications. I. Bouchet, Olivier. II. Title
TK5103.592.F73O6813   2005
621.382'7—dc22

2005033195

British Library Cataloguing-in-Publication Data
A CIP record for this book is available from the British Library
ISBN 10: 1-905209-02-9
ISBN 13: 978-1-905209-02-6

Printed in Brijbasi Art Press Ltd., I-72, Sector-9, Noida, U.P. India.

Table of Contents

Introduction

Communication methods employing optical transmission means existed, albeit in a very primitive form, for millennia; well before the work of the abbot Claude Chappe. However, the amount of information thus transmitted remained low. Telecommunications, which at first were exclusively optical, really started only at the end of the 18th century with the appearance of the optical telegraph. For fifty years, *wireless optics* enabled individuals to communicate over long distances. However, the quality of service (QoS) remained low due to the lack of reproducibility and reliability of both the transmitters and the receivers; of the men and the materials; and the changing nature of the air as a transmission medium.

Electricity (electric charges moving through copper) rapidly replaced optics (photons moving through the air). Information can be transferred along copper lines at relatively high flow rates. At the beginning of the third millennium, these copper-based links are still extensively exploited. Copper has served for decades as the basic material for very long distance links, and has been instrumental in establishing a network for the transmission of information across the whole globe.

The invention of the laser in 1960 paved the way to another solution, in the form of *optical telecommunication based on optical fibers*, which offers a quasi-unlimited line capacity. The almost simultaneous development in 1970–1971 of low-attenuation optical fibers, and semiconductor lasers emitting in continuous mode at room temperature, led to the explosion of wired optical telecommunications. Glass is the transmission medium for photons, and glass fibers can extend over distances of several thousand kilometers. *Wired optics* unquestionably currently dominates the fields of submarine transmissions, long-distance transmissions and interurban transmissions. It has become an integral and indispensable part of the Information Superhighway System.

For short distances, that is for the famous 'last kilometer' or 'last mile' as it is known in the telecommunication world, several different techniques, both wired and wireless, are currently competing: electricity with copper wire (xDSL, carrier current, etc.); fiber optics (glass or polymer); radio (GSM, UMTS, WiFi, WiMax, UWB, etc.); and now wireless optics. Each of these techniques presents advantages and disadvantages in terms of flows, transmission distances, costs and QoS.

Since the liberalization of the telecommunication sector, renewed interest has appeared in the digital transmission of signals in the atmosphere by laser beam. At a time when links between sites are multiplying, with an ever by increasing volume of information being transmitted, atmospheric optical links represent a wireless transmission mode with high flow rates (several Gbit/s) at short and average ranges (from a few decameters to a few kilometers). The principle of atmospheric links is the use of a wireless interconnection enabling communication between digital telephones, data-processing or video networks. This type of connection, which allows high information flow, is well-suited to short connections and, by extension, to networks with a limited dimensioning, for example wireless campus networks.

Several reasons can be offered for this revival of atmospheric optical links. Some of these reasons are of a legal nature: for instance, the exploitation of these links, in contrast to radio links, does not require any frequency authorization or any specific license. There are also economic reasons, since the deployment of a wireless link is easier, faster and less expensive for an operator than the engineering of optical cables. Finally, optics presents significant advantages compared to radio (even millimetric waves) in the race towards reaching flows of some Gbit/s. The availability of the components (receiving and modulating lasers, etc.), which are used extensively in telecommunication technologies based on optical fibers, further contributes to reducing equipment costs. Today, the world market for the wireless transmission of numerical data is primarily based on Hertzian technologies. However, these technologies present some limitations. In particular, due to their limited spectral width, it is unlikely that they alone will be capable of meeting the ever-increasing requirements in flows.

Throughout this book, the focus will be on the 1.5 μm wave. While the demonstrations and the reasoning can be applied to other photon wavelengths, our focus on the 1.5 μm wavelength is a deliberate choice, due to our belief that this wavelength will become the basic wavelength for wireless optical telecommunication systems with high flows. The advantages offered by this wavelength include a better ocular safety; greater availability of industrial components; the emission of photons at 1.5 μm using semiconductors or rare earth erbium; the possibility of communication systems operating in the continuation of other communication systems based on fibers (and therefore operating at 1.5 μm);

and less sensitivity to disturbances induced by the ambient light (sun, different illumination conditions, remote controls, etc.).

The objective of this book is to show how free-space optics, which is already commonly used for information transfer, is also taking off as a telecommunication technique, and is becoming an integral and essential part of data-processing architecture and telecommunications due its numerous advantages (flow, low cost, mobility of materials, safety, etc). First, a history of wireless optical telecommunications is presented, including the application of the principles of electromagnetism to free-space optics. Second, we describe transmitters and receivers of optical beams, which are the basis of any optical communication system: these devices were responsible for the first truly significant advances in the performances of these systems. Third, we devote special attention to the problems associated with the propagation of photons, both in the absence and in the presence of obstacles, since these are key issues for gaining an understanding of future telecommunication systems based on wireless optics. Finally, we consider standards as well as safety and confidentiality issues.

Chapter 1

History of Optical Telecommunications

1.1. Some definitions

1.1.1. *Telecommunication*

The definition of the word "telecommunication" adopted during the 1947 International Radiotelegraphic Conference held at Atlantic City (USA) is:

"any transmission, emission or reception of signs, signals, writings, images, sounds or information of any nature by wire, radioelectricity, optics or other electromagnetic systems."

The means of transmission must be electromagnetic in type, which gives a very wide scope, since, as Maxwell showed, electromagnetic waves include electricity and optics.

1.1.2. *Optical transmission*

This involves any transmission, emission or reception of visual signs and optical signals.

1.1.3. *Radio or Hertzian waves*

These are electromagnetic waves of frequency lower than 300 GHz; they are propagated in space without artificial guide (in optics, frequencies are significantly higher: hundreds of THz).

1.2. The prehistory of telecommunications

In the beginning, the need to communicate remotely was a natural reaction to life in a community. Communication must have been essential from the earliest times: even the days of Adam and Eve. The first men were already wireless operators, since they communicated with each other without wire, using sound and luminous waves (optical signals).

Homer mentions light signals in *The Iliad*: the fall of Troy was announced by fires lit on the hilltops. *The Anabasis* indicates this same mode of correspondence between Perseus and the army of Xerxes in Greece. In *The Agamemnon*, Aeschylus even gives details of the luminous relay stations used: eight relay stations were located on the tops of mountains such as Athos, Citheron and Aegiplanctus; covering around 550 kilometers. 400 years before Jesus Christ, the tactician Aeneas designed a code based on varying the number of torches. Two centuries later, Cleomenes and Polybius invented another code using a combination of light signals.

Figure 1.1: *This picture represents a Roman telegraph station, based on the low-relief of the Trojan column in Rome. Various signal systems have always been employed to transmit information quickly from one point to another...*

Later, the people of Carthage connected Africa to Sicily using very bright signals. After them, the Romans adapted the formula and all the Roman Empire was linked by fire signals (beacons) placed on watch towers (see, for example, Figure 1.1). Caesar made the optimum use of these during his campaigns in Western

Europe. Fires lit on high points, along the great Roman roads, were used to transmit rudimentary information of an essentially military nature quickly. Using such fires, lit from hill to hill, the Roman General Aetius forwarded the news of his victory over Attila to Rome in 451 AD.

During the siege of Nanking, the Chinese used kites with lamps to transmit signals in the same way. Similarly, the Native Americans used to transmit information using smoke signals.

Sailors have used semaphore and arm signals for a long time. The idea of alphabetical signals can be traced back to Ancient Greece, but it was in the Middle Ages that the first semaphore signals were used.

These communication processes, although very primitive, made use of optical transmission media. The transmission speed was suitable, but little information was contained in each message, because the various possible configurations of the light sources were very limited.

1.3. The optical air telegraph

Although the prehistory of telecommunication extends over millennia, the history of telecommunications really only starts at the end of the 18th century with the appearance, in France, of Claude Chappe's optical telegraph.

True semaphore, formed using wooden arms in a code corresponding to letters of the alphabet, seems to have been invented by the Englishman Robert Hooke (1684). His machine (see Figure 1.2) consisted of a broad screen which revealed letters or signals which corresponded to a code which had been agreed in advance. The system was improved by the French physicist Guillaume Amontons (1690). He suggested the use of a telescope to read the signals formed on the screen from a greater distance, in order to establish a correspondence, and thus communication, between two distant points.

Figure 1.2: *Robert Hooke's telegraph. The machine consists of a large screen (great width, great height) visible from a distance. Various signals can be seen. They indicate either letters of the alphabet or coded sentences agreed in advance*

During the whole of the 18th century, proposals for semaphore telegraphs followed one after another: Dupuis, Linguet, Courréjoles, and Brestrasser (arm signals) drew up various projects. But it was the system of Abbot Claude Chappe which was adopted by Convention in 1793, on the Lakanal's recommendation. In the ordinance dated September 24th, 1793, Chappe was given permission to use towers and bell-towers to install telegraph equipment between Paris and Lille. This connection, 230 km long, was built in record time between 1793 and 1794.

The Chappe telegraph machine was connected, in principle, to semaphore. At the top of the towers (Figure 1.3), in light of sight, was a system built of three articulated arms and driven by a mechanical device (Figure 1.4): each of the arms could take various positions. The set of the possible combinations, constituted a code that could be deciphered remotely thanks to the use of telescopes: the distance between the relay stations was, at most, only a few tens of kilometers.

Figure 1.3: *The Chappe system used high points: towers or bell-towers*

Figure 1.4: *Mechanism developed by Chappe: 3 articulated arms. The principal branch AB, horizontal in the figure, was approximately 4 meters long and 2 small branches called wings, AC and BC, were approximately 1 meter long. The mechanism is under the roof of the tower or bell-tower on which mast DD' is fixed. The mobile branches are cut out in the shape of shutters, which give the properties of great lightness, and wind resistance. The mobile branches are painted black to ensure good visibility against the sky background. Branch AB can give 4 positions: vertical, horizontal, right to left and left to right diagonal lines. Wings AC and BC can form right, acute or obtuse angles with branch AB. The figures thus formed are the signals. They are clear, easy to see, easy to copy down in writing, and it is impossible to confuse them. They allow a significant number of configurations*

The first great success of Chappe's telegraph was the announcement to the French Convention of the Landrecies victory (July 19th 1794). This involved transmission over a really long distance so this system represented considerable progress compared to the other media existing at the time. Chappe was appointed Telegraphs Director to reward this success. His invention was called the optical telegraph.

The Chappe telegraph thus entered history: however, it would only really take its place a few years later, when the second link was built on the strategic Paris–Strasbourg axis. This link, which included 50 stations, was brought into service in 1798, and the first dispatch was transmitted on July 1st 1798. It announced the capture of Malta by General Napoleon Bonaparte, during his Mediterranean Sea crossing towards Egypt. Next, a link was installed between Paris and Brest (1798), then another toward the South of France (1799). The Paris–Lyon link was installed in 1806 and was extended towards Turin and Milan (1809). The Northern link reached Antwerp, via Brussels, in 1809.

The Louis-Philippe government undertook the construction of a network designed on the pattern of a spider's web, with Paris at its centre in order to improve the efficiency of its administration. This network supplemented the previously installed links between Paris and the larger French cities (Lille, Brest, Strasbourg), then installed concentric links, centered on Paris, and recutting the radiant links from place to place. This network offered alternative ways to transmit telegrams if communications with Paris were cut. After this, the network was enlarged by linking the Northern fortified towns, the commercial areas around the coast, and the major cities of the South of France. Thus, in 1844, France had a network of 534 semaphore stations covering more than 5000 km. This Chappe air network had a characteristic in common with all the French communication networks, in particular, roads and railways: it was built on a star topology centered on Paris.

However, after 1845, the electric telegraph appeared and gradually it took the place of the optical telegraph. The last link of the free-space telegraph network was taken out of the network in 1859.

The Chappe invention was also very much exploited in other countries, especially Spain and Italy. In Russia, Tsar Nicholas I established links between Moscow, St Petersburg, Warsaw and Cromstadt. He inaugurated the Moscow–Warsaw links in 1838. These links comprised 200 stations served by 1320 operators. In the most difficult areas, fog made the lengthened signals hard to see. The system was adapted to make use of mobile shutters giving combinations which were varied enough to offer a multitude of signals. Free-space telegraphs were also built according to this system in England and Sweden.

1.4. The code

To have an operating telegraph, Chappe had to solve two problems: concerning the mechanism and the code. To be effective, the code was not alphabetical: the transmission of the shortest sentence would have required as many signals as letters, which would have taken a considerable time. Thus Chappe adopted a set of codes: signals or rather groups of signals were linked to words, sentences or geographical names. Each signal was made using three articulated elements. A characteristic geometrical figure was created by varying the angle of each element (Figure 1.5). These shapes could be identified from a distance. The central arm was either horizontal or vertical and the two wings could be horizontal, vertical or left or right diagonal.

"Let us delay a little on this invention which introduced the concept of the information and telecommunications network. A link was effected by a succession of towers, at the top of which a mechanism formed of articulated arms was installed whose various positions, 92 in all, each corresponded to a number. Ethnologists have since highlighted that ants exchange information thanks to this type of signal. In each relay station tower, an operator playing the role of repeater-regenerator reproduced, without understanding the significance, the code dispatched to him by the preceding tower, so that it could be observed with a telescope. The operator of the next station transcribed the message received on paper, in the form of a succession of numbers, and carried it to the addressee who, using a highly confidential coding document, possessed by only a few, associated a word to each number and thus reconstituted the message. This system simultaneously combined digital transmission; the encoding intended to protect information; and error correction. The numerical data rate was low since it took approximately 20 seconds to transmit a combination, which corresponds to approximately 0.4 bit/s, but the Chappe optical telegraph brought an extraordinary gain to the transmission speed compared to the best method known hitherto, the transport of the mail by a rider. The first message was transmitted by this technique on March 2nd 1791 and a veritable network was constituted little by little: it would continue to be used until around 1855, when it would be supplanted by the electric telegraph, which offered the advantages of a higher transmission capacity and 24/7 usage, including during the night as well as during times of low visibility.

Thus the term 'optical' disappeared for a time from the transmissions field."

Box 1.1: *The Chappe telegraph [JOINDOT, 1996]*

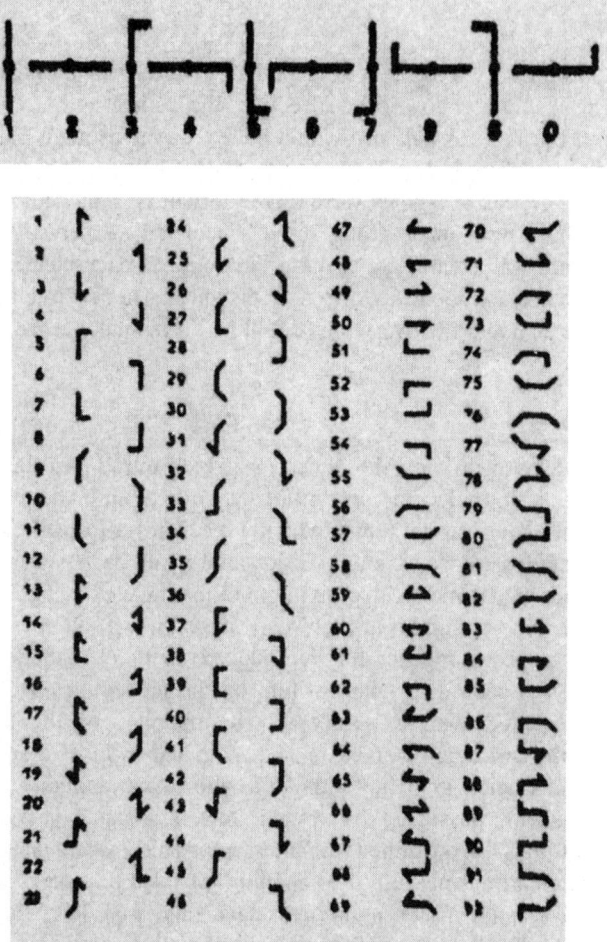

Figure 1.5: *Signals used in the final code of the Chappe telegraph. The code of the Chappe telegraph evolved over the course of time. However, it should be noted that, from the beginning, the Chappe telegraph did not use an alpha-digital code, but a code similar to that of diplomats: the idea came from a relative of the inventor Leon Delaunay, former French Consul in Lisbon. The signs transmitted using the arms of the apparatus corresponded to figures or simple numbers. At the beginning, the number of the signals was limited to ten; the articulated arms could only take horizontal or vertical positions (see the top of the figure). Then, in about 1800, their number increased from 10 to 88, then to 92. The first code used by Chappe was composed of 9999 words, each one represented by a number. Later, there were three codes: a "word" code, a "sentence" code and a "geographical" code. These three codes were amalgamated into one code in 1830: this new code is undoubtedly that which is preserved at the Post office museum at Paris [LIBOIS, 1994]*

The "vocabulary" of this code is contained in an imposing register of 704 pages (88*4+88*4), comprising some 45000 terms, sentences or geographical names. In theory, the transmission was done by group of two numbers; the first corresponded to the vocabulary page and the second to the text line.

Thus, one could have, for example, under the "army" heading:
*page 36, line 6: "**armée**"*
*page 35, line 9: "**armée ennemie composée de**"*
*page 35, line 13: "**l'armée a battu complètement l'ennemi**"*

and, under the heading "dépêche":
*page 53, line 12: "**dépêche de la plus haute importance**"*
*page 53, line 13: "**la fin de cette dépêche ne m'est pas encore parvenue**"*
*page 53, line 21: "**je réponds à votre dernière dépêche**"*

Thus a text such as:
*"**je réponds à votre dernière dépêche; l'armée a battu complètement l'ennemi**" required only the two groups: 53 – 21 and 35 – 13*

Such a system ensured the secrecy of the messages and economy of signals at the same time. It also explains the stereotyped style of the dispatches sent by this means during the time of the French Revolution and the French Empire.

Box 1.2: *The Chappe Code [Libois, unpublished]*

1.5. The optical telegraph

The Chappe free-space optical telegraph described earlier was quickly and completely replaced by the electrical telegraph. For the optical communication network to be used again, it would need to adapt and find new uses. In fact, it was transformed, and became the new optical telegraph. The new optical telegraph could render multiple services through the transmission of messages in particular circumstances, such as transmissions between army corps during war or peace time. Like the Chappe telegraph, this new optical telegraph produced signals visible at long distances, using telescopes. However, these signals were composed of flashes of sunlight or oil lamps, corresponding to Morse code characters. This system, developed by Lesuerre, first drew attention during the siege of Paris in 1870–1871. Whereas electrical telegraphy systems had been ransacked and thus become inoperative, it was thanks to the optical transmission system that correspondence could continue despite everything which was done between the forts surrounding Paris. It was fortunate then to have a communication system which passed over the heads of the enemy.

Mangin proposed a clearly improved system. The source was an oil lamp which used a set of mirrors and a shutter which gave a flashed parallel light beam. The longer or shorted flashes corresponded to dashes and dots: features of the Morse Code telegraph. A fixed telescope set parallel to the axis of the lenses was used to see the signals of the transmitting station. The oil lamp could be replaced by an optical system using sunlight (Figure 1.6, according to Max de Nansouty, 1911) [De NANSOUTY, 1911].

The speed these luminous flashes can be emitted is about twenty words per minute. Using sunlight during the day and an oil lamp during the night, the optical lenses apparatus allowed communication for distances from 30 km up to 120 km. The signals were seen using a telescope joined to the receiver apparatus. Maximum range transmission required absolutely clear air, because light signals are absorbed by fog, smoke and haze.

Figure 1.6: *Mangin's optical telegraph. At the top, the light source is an oil lamp and the shutter A is used to order the light signal emission (points and features of the Morse telegraph), telescope EE' is parallel to the axis of the lenses and used to see the signals of the transmitting station; at the bottom, the telegraph is lit by the sunlight during the day time, the two plane mirrors M and M' are tilted according to the position of the sun to return the pencil of light refracted in the apparatus and through a set of lenses, one thus has an output pencil of light as in the preceding case*

1.6. The heliograph or solar telegraph: a portable telecommunication system

The heliograph or solar telegraph is a portable optical telecommunication system which can transmit signals for distances of up to 40 km. This uses the reflection of solar rays with a directional mirror. The reflected rays are directed towards the receiving station (Figures 1.7 and 1.8, Max de Nansouty, "Electricity", 1911). Contrary to the other types of apparatus considered earlier, which were based on the controlled emission of light signals, this telegraph emits signals by extinction of luminous flux (at rest it always lights the transmitting station.) The major disadvantage is that it only functions during daytime, because the sunlight is its only transmission support.

Figure 1.7: *English heliograph: when one wants to transmit signals, one directs the mirror M whose reflected rays will cross the test card D; this system does not function if the sun is high on the horizon*

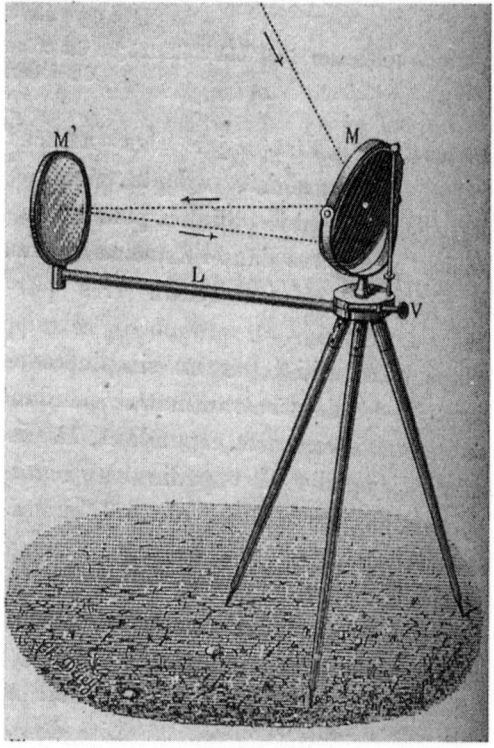

Figure 1.8: *Heliograph with double mirror. The use of a second mirror allows the collection of sun rays, falling on the first with a very open angle and reflecting onto the main mirror. This double mirror system allows use on a much broader timetable than the preceding system*

1.7. Alexander Graham Bell's photophone

In 1880, four years after having invented the telephone, Alexander Graham Bell made his first wireless optical communication in which the rays of the sun replaced electric wire:

"I heard the sun rays to laugh, to cough and to sing," he wrote to his father after the first demonstration of the operation of this apparatus, which was to be named the "photophone". It was a transmission in free space over a length of about 200 meters, transmission using the sunlight as carrier and thus showing, for the first time, the basic principle of modern optical communications where the optical fiber replaces the air as the medium for transmission of light.

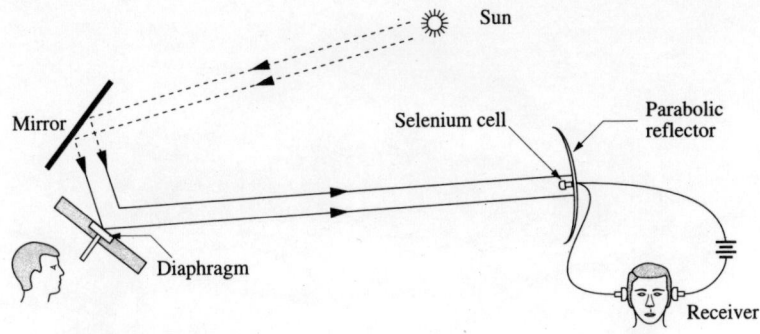

Figure 1.9: *Schematized representation of the photophone invented by Graham Bell; the sunlight is modulated by a vibrating diaphragm and is transmitted in free space about 200 meters, then delivered by a selenium cell connected to ear-phones (source: http://cord.org/step-online/st1-7/st17ei.htm)*

The basic principle of the photophone is schematized in Figure 1.9. The sunlight is focused onto a flexible reflective membrane. The user speaks into this membrane. The spoken word is transmitted in the air by modulating the reflected sun rays. This modulated reflected light, after its displacement in the air, is collected by a photoconductive selenium cell connected to a pile and the ear-phones.

Graham Bell considered that this invention, patent deposited on June 3rd 1880, was his greater invention, much more significant than the telephone. And yet we all know what happened. The transmission lengths were too small and the sun does not emit in exactly the same manner 24 hours a day, 365 days a year.

However, 120 years later, the engineers took up the idea of the photophone and used free-space optical links for wireless transmissions at very high data rate, as suggested by Figure 1.10. The sun, definitely too variable, could not be used, but perfectly controllable pencils of light were used instead, thanks to optical telecommunication in optical fibers. These beams of photons are provided by lasers whose characteristics are known exactly (wavelength, number of emitted photons, time or sequences of emission); these are collected by extremely sensitive and reproducible detectors. The medium of transmission remains the air and can be variable (for example, due to changes in climate, vegetation, industrial environment), but optics and modern electronics can mitigate a degree of inhomogeneity. This problem of non reproducibility of the transmission medium is the same one as that met for communications in radioelectricity such as in broadcasting. It can be solved by using suitable electronics in the receivers.

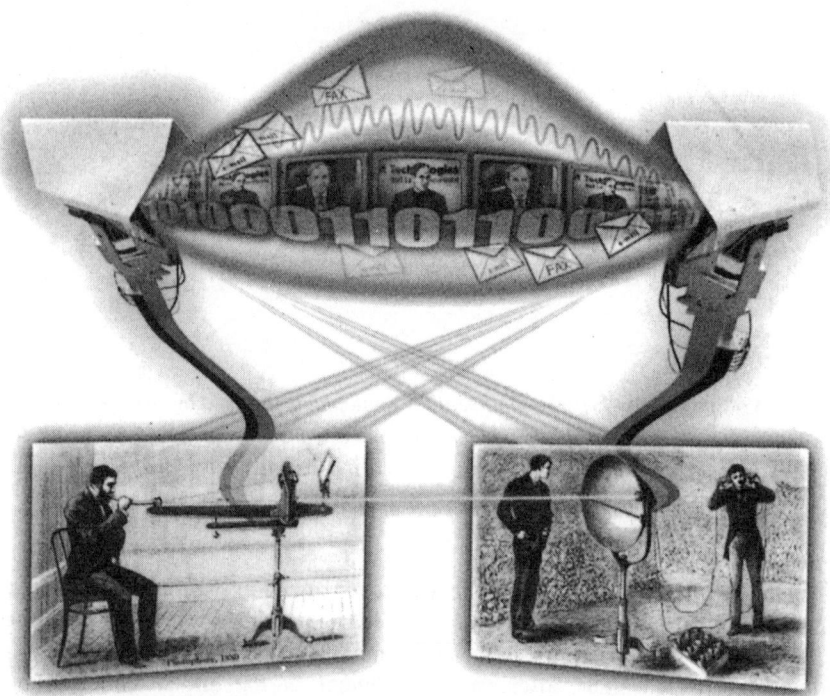

Figure 1.10: *In 1880, Graham Bell transmitted his voice over more than 600 feet (200m), through the air and using the reflected rays of the sun. In 1999, engineers took up the idea of the photophone*
(source: http://www.bell-labs.com/news/1999/february/ 25/presby_lg.jpeg)

Chapter 2

Basic Principles of Electromagnetism

2.1. Introduction

Information can be transmitted in various ways. Among the possibilities, one of the more powerful is the use of electromagnetic waves for the transfer of information. Let us return to these in more detail here. This chapter gives a traditional description of electromagnetic waves. We shall introduce Maxwell's equations for the propagation of electromagnetic waves in various media, and give the expression for the energy associated with these waves. Finally, we will present models for the propagation of rays.

2.2. Maxwell's equations in an unspecified medium

The propagation of electromagnetic waves is deduced from Maxwell's equations [BRUHAT, 1992, COZANNET, 1983, VASSALLO, 1980, SEELY, 1979, BORN, 1983].

$$\overrightarrow{\mathrm{rot}}\ \vec{E} = -\frac{\partial \vec{B}}{\partial t} \tag{2.1}$$

$$\overrightarrow{\mathrm{rot}}\ \vec{H} = \vec{j} + \frac{\partial \vec{D}}{\partial t} \tag{2.2}$$

$$\mathrm{div}\,\vec{B} = 0 \tag{2.3}$$

$$\mathrm{div}\,\vec{D} = \rho \tag{2.4}$$

$\vec{E}$ is the electric field (in V/m), and $\vec{H}$ the magnetic field (expressed in A/m) associated with the electromagnetic wave. $\vec{D}$ and $\vec{B}$ are the electric displacement (in C/m^2) and magnetic induction (in Wb/m^2 or T), respectively, which describe the influence of a medium on the propagation of electromagnetic waves. ρ is the electric charge density (in C/m^3) and $\vec{j}$ is the electric current density (in A/m^2). $\vec{j}$ and ρ are bound by the condition of charge conservation, that is, charge is conserved at any point:

$$\text{div } \vec{j} + \frac{\partial \rho}{\partial t} = 0 \qquad [2.5]$$

Using these equations, it is possible to determine the electromagnetic wave propagation in any medium. In the case of free-space propagation, which is the subject of this book, the set of preceding equations is simplified. The wave propagates in a homogeneous medium which contains no electric charges or currents (i.e. $\rho = 0$ and $\vec{j} = \vec{0}$). Moreover, the transmission medium is assumed to be isotropic, non dispersive and linear, so:

$$\vec{D} = \varepsilon \vec{E} \qquad [2.6]$$

$$\vec{B} = \mu \vec{H} \qquad [2.7]$$

where ε and μ are the permittivity (or dielectric constant) and the permeability (or magnetic constant), respectively.

The simplest case is propagation in a vacuum, where $\varepsilon = \varepsilon_0 = 1/(36\,\pi 10^{-9})$ (in F/m) and $\mu = \mu_0 = 4\pi\,10^{-7}$ (in H/m). For propagation in a medium, $\varepsilon/\varepsilon_0$ is the relative permittivity ε_R and μ/μ_0 the relative permeability μ_R.

To solve this problem, the sinusoidal mode of time is considered. In other words, one searches for solutions existing between $t = -\infty$ and $t = +\infty$. It is also assumed that the fields and inductions vary sinusoidally with time, with the same angular frequency ω. For example, the real electric field $\vec{E}_r$ will be written as:

$$\vec{E}_r = \vec{E}_0 \cos(\omega t + \varphi)$$

In order to simplify the calculations, exponential notation will generally be preferred:

$$\vec{E}_c = \vec{E}_0 \exp(i\phi)$$

is deduced from $\vec{E}_c$ using the equation:

$$\vec{E}_r = \text{Re}\{\vec{E}_c \exp(i\omega t)\} = \frac{1}{2}\{\vec{E}_c \exp(i\omega t) + \vec{E}_c^* \exp(-i\omega t)\}$$

This notation will be used throughout this chapter, and the field $\vec{E}_c$ will be written simply as $\vec{E}$.

2.3. Propagation of electromagnetic waves in an isotropic and linear homogeneous medium

The goal of this part is to characterize the waves propagating in a linear and isotropic homogeneous medium. The medium is homogeneous if the permittivity ε is constant. Let us consider the simplest case, i.e. that of propagation in a homogeneous medium without any electrical charge. Using the complex notation, $\vec{E}$ and $\vec{H}$ can be written as:

$$\overrightarrow{\text{rot}}\, \vec{E} = -i\omega\mu\vec{H}, \tag{2.8}$$

$$\overrightarrow{\text{rot}}\, \vec{H} = i\omega\varepsilon\vec{E}, \tag{2.9}$$

$$\text{div}\, \vec{H} = 0, \tag{2.10}$$

$$\text{div}\, \vec{E} = 0. \tag{2.11}$$

From the four preceding relations, the two following equations are deduced:

$$\Delta\vec{E} + \omega^2 \varepsilon\mu_0 \vec{E} = \vec{0} \tag{2.12}$$

$$\Delta\vec{H} + \omega^2 \varepsilon\mu_0 \vec{H} = \vec{0} \tag{2.13}$$

where Δ is the Laplacian operator.

These two relations are generally known as the equations of propagation. Their analytical resolution shows that:

$$\vec{E} = \vec{E}_0 \exp(-i\vec{k} \cdot \vec{r}) \tag{2.14}$$

and

$$\vec{H} = \vec{H}_0 \exp(-i\vec{k} \cdot \vec{r}) \qquad\qquad [2.15]$$

with $\left|\vec{k}\right|^2 = k^2 = \omega^2 \varepsilon \mu_0$.

$\vec{k}$ is the wave vector or propagation vector. It expresses the spatial frequency of the wave. The spatial period λ is connected to $\vec{k}$ through: $\left|\vec{k}\right| = 2\pi/\lambda$.

$\vec{E}, \vec{k}, \vec{H}$ form a right-handed orthogonal trihedral of vectors. In other words, if the electric field $\vec{E}$ propagates along the direction given by its wave vector $\vec{k}$, the associated magnetic field $\vec{H}$ has a single direction. The real fields $\vec{E}_r$ and $\vec{H}_r$ vary sinusoidally with time. Their dependence on the distance is shown in Figure 2.1. For a propagation in a homogeneous, dielectric and isotropic material, $\vec{E}_r$ and $\vec{H}_r$, as well as their associated complex notations, lie in a plane normal to the direction of propagation. In this case, the electromagnetic vibration is said to be transverse.

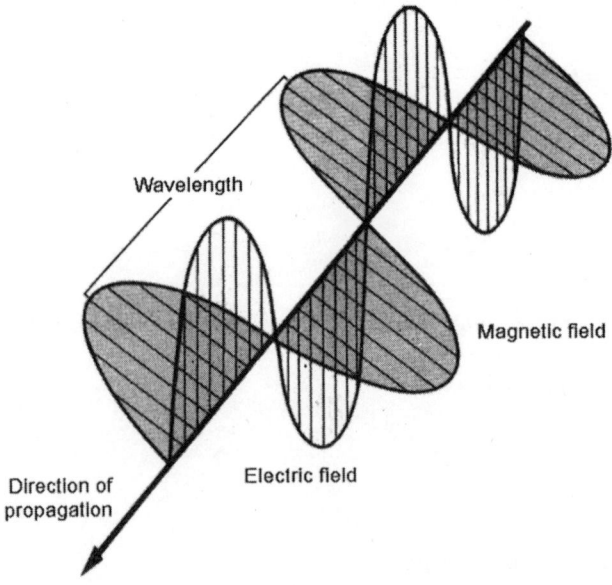

Figure 2.1: *Representation of the real components of the electric and magnetic fields along the direction of propagation*

From the previous relations, the phase of the wave can be written in the form:

$$\phi(\vec{k},\vec{r}) = \omega t - \vec{k}\cdot\vec{r} \tag{2.16}$$

The phase velocity, i.e. the speed necessary for remaining at a constant phase is:

$$v_\phi = \frac{\omega}{k} \tag{2.17}$$

For a wave propagating in a medium:

$$v_\phi = \frac{c}{n},$$

where c is the velocity of light in a vacuum and n is the refractive index of the medium.

2.4. Energy associated with a wave

The energy balance is one of the most significant parameters for the link. Until now, we have considered plane waves extending in all the planes normal to the direction of propagation. For optical links, the beams will be limited laterally. We shall recall here the expression for luminous flux through surfaces in the case of plane waves.

In order to calculate the energy carried by a wave, let us first recall that the electromagnetic energy present in a volume V is expressed by the relation [BRUHAT, 1992]:

$$U = \int_V (\varepsilon|\vec{E}|^2 + \mu_0|\vec{H}|^2)dV \tag{2.18}$$

The energy density is given by:

$$w_{em} = \frac{1}{4}(\varepsilon|\vec{E}|^2 + \mu_0|\vec{H}|^2)$$

In the case of a plane wave propagating in a perfect medium, it can be shown that the electrostatic energy is equal to the magnetic energy. The electromagnetic energy density is therefore:

$$w_{em} = \frac{1}{2}\varepsilon|\vec{E}|^2 \tag{2.19}$$

Let us now consider the luminous flux through an uncharged homogeneous surface, normal to the direction of propagation, where the normal to the surface is parallel to the z-axis. Let us consider the amount of energy during time $dt = dz/v$, where I is the speed of light in the medium considered. If the electric and magnetic fields are linearly polarized along Ox and Oy, it can be shown that the energy flux through the plane is equal to the flux of a normal vector $\vec{P}$ given by:

$$\vec{P} = \frac{1}{2}\vec{E} \wedge \vec{H}^*$$
[2.20]

The component along z of the Poynting vector is written as [BRUHAT, 1992]:

$$P_z = \frac{dU}{dz} = \frac{1}{2}E_x H_y^*$$
[2.21]

The Poynting vector is associated with the energy flux with respect to both its magnitude and its direction of propagation.

The instantaneous energy associated with the wave cannot be measured due to the fact that the frequency of light is about 10^{15} Hz, and the detectors measure only the average energy. Indeed, only the energy average over a period T is observable:

$$\langle \vec{P} \rangle = \frac{1}{T}\int_0^T \vec{P}dt = \frac{1}{2}\text{Re}(\vec{E} \wedge \vec{H}^*)$$
[2.22]

The average flux of the Poynting vector through a surface corresponds to the average energy carried by the wave through this surface.

Note – here again, it is necessary to be careful depending on whether the real or complex notation of the electromagnetic fields is used [VASSALLO, 1980].

2.5. Propagation of a wave in a non-homogeneous medium

The plane waves described previously give the ideal case of an infinite, homogeneous, linear and lossless medium. The case of a wave propagating in the air comes close to this ideal case. Of course, a part of electromagnetic energy transmitted in the atmosphere will be absorbed by the molecules present in the air. The absorption processes will be described in more detail in the following chapter. In a bounded medium, as is the case in an optical fiber or waveguide, it is necessary to take Maxwell's equations regarding the differences between the media into account. The equations of propagation will no longer assume the simple form given by equations [2.14] and [2.15]. Since a purely analytical solution is not always available, it is necessary to use numerical algorithms for the resolution of the propagation equations of an electromagnetic wave in the system under consideration. Let us mention some algorithms such as the FDTD, the BPM and the finite element method.

2.6. Coherent and incoherent waves

In free-space propagation, the diffraction and reflection processes generate several waves which interfere with one another (Figure 2.2). If two waves have a phase relation, i.e. are coherent, the fields will be summed in amplitude and phase. The resulting field depends on the phase difference between the waves. If the two waves are incoherent with respect to one another, i.e. if there is no phase relation, the intensities still sum up. Thus it is important to consider the phase relation between the detected signals.

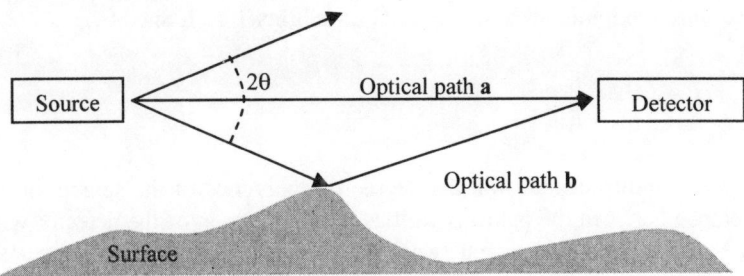

Figure 2.2: *Diagram showing the summing up of two light beams, one propagating in straight line (a) and the other reflected at an irregularity on the surface (b), 2θ is the numerical aperture of the source*

There is a more complete account of coherence and interference phenomena in the references given in the preceding section. Here, we shall only mention a few properties defined, for plane waves, as well as some definitions and relations associated with the coherence of different detected signals. It has been shown that the density of electromagnetic energy is proportional to the square of the electric field. Let us consider the fields E_a and E_b emitted by the source after their propagation along the paths a and b with lengths L_a and L_b respectively:

$$\vec{E}_{a,b} = \vec{E}_0 \exp(-jkl_{a,b})$$
[2.23]

The total field can be written in the form:

$$\vec{E}_{tot} = \vec{E}_0 \exp(-jkl_a) + \vec{E}_0 \exp(-jkl_b)$$
[2.24]

The average intensity of the field at the detector is therefore:

$$I = |\vec{E}|^2 = 2|\vec{E}_0|^2 [1 + \cos k(l_b - l_a)] \qquad [2.25]$$

Variations in the signal can be observed due to the phase difference between the beams. The ideal case corresponds to that of perfectly monochromatic waves, where the wave emission of the source is infinite. Reality can be very different: the term including the phase difference can be zero. In this case the two beams are said to be incoherent with respect to one another, and therefore their respective intensities will be directly added. The reduction of the coherence of a source can be associated either with the space incoherence of the source related to its spatial extension or with the spectral width of the source which is not null as in the case of an ideal monochromatic light.

When the length difference between the two optical paths varies, the energy passes through minima and maxima. The visibility V is defined by:

$$V = \frac{I_{Max} - I_{min}}{I_{Max} + I_{min}} \qquad [2.26]$$

The visibility depends on the degree of coherence of the source, on the length difference between the paths as well as on the location of the detector with respect to the source. The coherence between the various beams arriving at the detector also depends on the crossed media: for example the diffusing medium can reduce the coherence. For links referred to as "in direct sight" links, coherent sources can be used, provided that parasitic reflections do not interfere with the principal beam, inducing modulations of the detected signal.

Lasers have a coherence length longer than a meter, whereas the coherence length of white light sources can be of the order of the micron.

2.7. Relations between classical electromagnetism and geometrical optics

In the remainder of this book, geometrical optics will be extensively used. First, we shall briefly outline the transition from electromagnetism to geometrical optics and explain why the use of ray optics is fully justified in this context. Let us consider the case of an actual non-homogeneous medium. The refractive index can vary. Thus we search for solutions of the following form:

$$\vec{E} = \vec{E}_0(r)e\exp[-ik_0 S(\vec{r})],$$
$$\vec{H} = \vec{H}_0(r)\exp[-ik_0 S(\vec{r})]$$

[2.27]

with $k_0 = \dfrac{2\pi}{\lambda} = \dfrac{\omega}{c}$.

This wave is a harmonic wave. The quantity $S(\vec{r})$ is called *Eikonal*, $S(\vec{r})$ is comparable to a distance and corresponds to the optical path, while $\vec{E}_0(r)$ and $\vec{H}_0(r)$ are vector quantities. The readers interested in the *Eikonal* function are referred to the two following references: [BORN, 1983] and [COZANNET, 1983]. Here we briefly describe some of their properties, as well as the relation between rays and waves.

Using Maxwell's equations, equations [2.8] to [2.11], and with the notation $\vec{k}(\vec{r}) = k_0 \overrightarrow{grad}(S)$ in the equations above, one obtains:

$$\vec{k} \wedge \vec{E}_0 - \omega\mu\vec{H}_0 = -i\overrightarrow{rot}\vec{E}_0$$
$$\vec{k} \wedge \vec{H}_0 - \omega\varepsilon\vec{E}_0 = -i\overrightarrow{rot}\vec{H}_0$$

[2.28]

Of course, if $\vec{E}_0(r)$ and $\vec{H}_0(r)$ were constant, one would find the same relations as in the case of a plane wave. The plane wave approximation is justified, provided that:

$$\frac{\Delta E}{E} << \frac{\Delta x}{\lambda_0}$$
$$\frac{\Delta H}{H} << \frac{\Delta x}{\lambda_0}$$

The harmonic wave thus turns out to be similar to a plane wave. In other words, if the variations in amplitude of the fields are small with respect to the wavelength, the harmonic wave can be considered locally as a plane wave.

The wavefront is the locus of the points which are in phase, i.e. where $S(\vec{r})$ is constant. For a plane wave, the wavefront is a plane. For a harmonic wave, it may assume any specific shape (see Figure 2.3). The normal to the wavefront of the harmonic wave is given by the vector $\vec{n}_f$, which of course depends on the point considered in the wavefront.

$$\vec{n}_f = \frac{\overrightarrow{gradS}}{\left|\overrightarrow{gradS}\right|}$$

[2.29]

For any point, the vector $\overrightarrow{gradS}$ defines a normal to the wavefront associated with this point. This corresponds to the definition of a light ray in geometrical optics. Likewise, $gradS$ is related to the refractive index of the medium where the harmonic wave propagates through the equation:

$$\left|\overrightarrow{gradS}\right|^2 = c^2 \varepsilon \mu = n^2 \, .$$

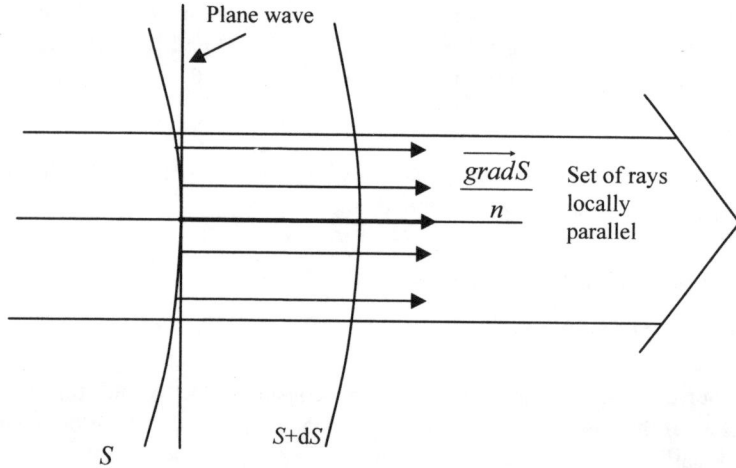

Figure 2.3: *Local approximation of a wave surface to a plane wave. Each element of the surface is compared to its tangent plane. The normal to this wavefront defines a light ray*

The concept of plane wave, like that of light ray, is a purely theoretical notion used for approaching reality in a simplified way. The concept of light ray assumes a point source coming from infinity, while the concept of plane wave supposes a plane wave extending to any point of space. This means that in the following, every time the notion of light ray is used, it is implicitly assumed that a plane wave can be defined locally.

2.8. The electromagnetic spectrum

Different bands of frequency can be associated with electromagnetic waves. These bands, while being comparable in nature, have different names for historical reasons. This is especially true in optics, which is concerned with frequencies corresponding to visible light, whereas the electromagnetic spectrum largely extends above and beyond this. The spectrum is also divided into two regions: ionizing frequency bands and non-ionizing frequency bands, depending on the impact of these waves on biological tissues. Ionization occurs when an electron is withdrawn from its normal location in the atom or the molecule, causing damage to biological tissues. The ionizing part of the electromagnetic spectrum includes ultraviolet rays, gamma-rays and X-rays, whose wavelengths are very short, beginning from 400 nanometers, and whose intensities are very high. The non-ionizing part includes radio waves, microwaves, infra-red and visible light. Figure 2.4 shows the electromagnetic spectrum extending with respect to the wavelength (in meters) with the usual names for the various frequency bands. An enlarged view is represented for wavelengths in the band of the visible light. For a detailed definition of the different frequency bands, see for example Sizun's book [SIZUN, 2003].

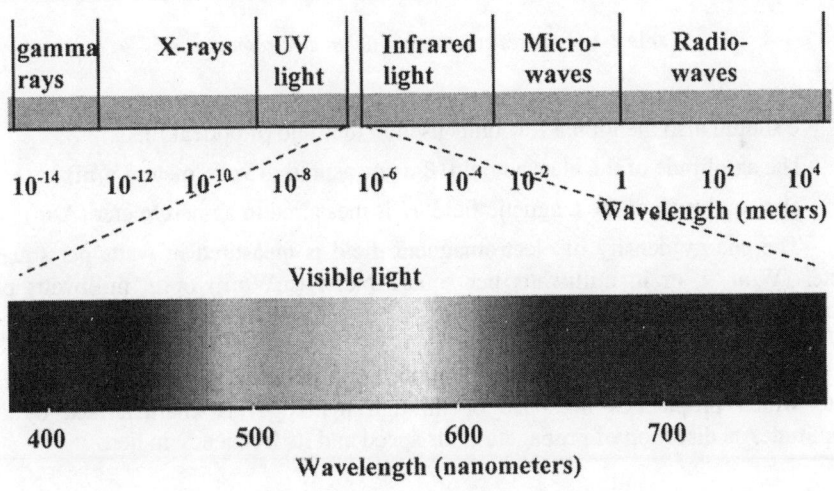

Figure 2.4: *The electromagnetic spectrum with respect to the wavelength with the usual names for the various frequency bands*

2.9. Units and scales

As can be seen, the spectrum of electromagnetic waves is very wide, with wavelengths ranging from 10^{-14} up to 10^4 m. Both very long and very short wavelengths will be encountered throughout this book. The table below (Table 2-1) gives the meaning of some of the prefixes that will be used on these occasions.

Prefix	Symbol	Unit	Nominal Value	Numerical Value
Tera-	T	10^{12}	One trillion	× 1,000,000,000,000
Giga-	G	10^9	One billion	× 1,000,000,000
Mega-	M	10^6	One million	× 1,000,000
kilo-	k	10^3	One thousand	× 1,000
milli-	m	10^{-3}	One-thousandth	× 0.001
micro-	μ	10^{-6}	One-millionth	× 0.000001
nano-	n	10^{-9}	One-billionth	× 0.000000001
pico-	p	10^{-12}	One-trillionth	× 0.000000000001
femto-	f	10^{-15}	One-quadrillionth	× 0.000000000000001
atto-	a	10^{-18}	One-quintillionth	× 0.000000000000000001

Table 2-1: *Prefixes commonly used in electromagnetism*

We should also mention a few units used in the field of optical links:

– The amplitude of the electric field $\vec{E}$ is measured in volts/meter (V/m).

– The amplitude of the magnetic field $\vec{H}$ is measured in ampere/meter (A/m).

– The energy density of electromagnetic field is measured in watts per square meter (W/m ^{2}), or in milliwatts per square meter (mW/m²) or in milliwatts per square centimeter (mW/cm²).

Fundamentally, a wave is an oscillation, i.e. a periodic variation of a physical state which propagates in space or through matter. It is characterized by its amplitude, its direction of propagation, its speed and its frequency in hertz.

The number of oscillations per unit of time is referred to as the frequency, and is measured in hertz (Hz).

The time interval between two successive oscillations with same direction and size is called the period. Its unit is the second (s).

The space traveled in by the wave during this time is the wavelength. Its unit is the meter (m).

The set of points reached by the disturbance in a homogeneous medium after a given time, starting from the time of emission, is referred to as the wave surface or wavefront.

The relation between these different units (frequency, period, wavelength) is presented below:

– the frequency $f = 1/T$

   - f = frequency in hertz (or multiple)
   - T = period in seconds (or multiple)

– the wavelength $\lambda = c \times T = c/f$

   - T = period in seconds (or multiple)
   - c = velocity or speed of light = 3×10^8 m/s or 299 792 km/s
   - f = frequency in hertz (or multiple)

– the period $T = 1/f$

   - T = period in seconds (or multiple)
   - f = frequency in hertz (or multiple)

Depending on the field of study, different units are used for the various spectral bands. Table 2-2 shows the relations between the frequency, the period and the wavelength, and the characteristics, depending on the part of the electromagnetic spectrum being considered. Different parameters are generally chosen merely for ease of use, but are associated with the same physical phenomenon.

Waves	Wavelength	Frequency	Period	Characteristics
Radiowaves	100 km to 1 m	10 kHz to 300 MHz	100 ns to 0.1 ms	Radio FM, AM and TV
Microwaves	1 m to 1 mm	300 MHz to 300 GHz	3 ps to 100 ns	Mobile, satellite, radar
Infrared	1 mm to 0.8 μm	300 GHz to 300 THz	3 fs to 3 ps	Laser, night sight, telemeter
Visible light	400 to 800 nm	Energy in electron-volts (eV) – 1 to 5 eV	1 to 3 fs	Laser, sun, lamps
Ultraviolet	400 to 0.5 nm	5 to 1 keV	1.7as to 1 fs	Laser, lamps
X-rays	50 to 0.1 pm	1 to 100 eV	0.0003 to 1.7 as	X-rays tubes
Gamma-rays	< 0.1 pm	> 100 eV	< 0.0003 as	Radiation of energetic particles, synchrotrons

Table 2-2: *Table showing the frequencies, periods and corresponding wavelengths*

2.10. Examples of sources in the visible light and near visible light

We shall now present a few examples of optical sources, and provide some information concerning solar radiation. Indeed, it is important to know the spectral signature of these sources of perturbation for optical links since these radiations may disturb optical links in free space [FREEMAN, 1960].

The main natural source of electromagnetic radiation is the Sun. Natural electromagnetic energy (solar radiation) allows, amongst other things, the photosynthesis of trees and plants. Its spectrum extends from 300 nm to more than 1500 nm, with varying intensities or amplitudes. The peak intensity is located around 480 nm (corresponding to the color blue) before progressively decreasing as the wavelength increases. Our eye perceives but a small fraction of solar radiation, that between 400 and 700 nm (see Figure 2.4). The Sun emits incoherent radiation; the Sun's spectrum is shown in Figure 2.5.

The Sun can be approximated as a black body. Another example of a black body is a filament lamp. The intensity of radiation from a black body at temperature T is given by Planck's law of black body radiation:

$$I(v) = \frac{2hv^3}{c^2} \frac{1}{\exp(hv/kT)-1} \qquad [2.30]$$

where T is the temperature of the element (in K), h is Planck's constant, k is the Boltzmann constant and c is the velocity of the light. For the Sun, this spectrum exhibits a maximum at the frequency $f = 4 \times 10^{14}$ Hz and the corresponding bandwidth around 4×10^{14} Hz.

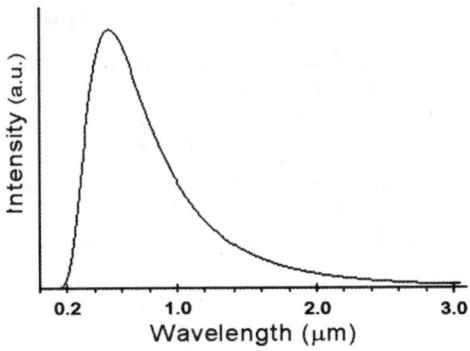

Figure 2.5: *Curve of brightness of solar radiation. The absorption peaks of water, ozone, and oxygen molecules are not represented for the sake of clarity*

Another source of electromagnetic radiation, commonly used in both office and home environments, is the tungsten filament lamp. Figure 2.6 shows the emission spectrum of a 60 W lamp. This spectrum is very broad and continuous, and exhibits a peak around 1000 nm. It might be stressed that optical communications use sources with a quasi-monochromatic low bandwidth (Figure 2.7) whose spectra can be compared instructively with the spectra resulting from ambient lights (Sun, lamps etc.). However, it is clear that for optical communication in free space, any source emitting in the spectral window of the detector is likely to disturb the transfer of information.

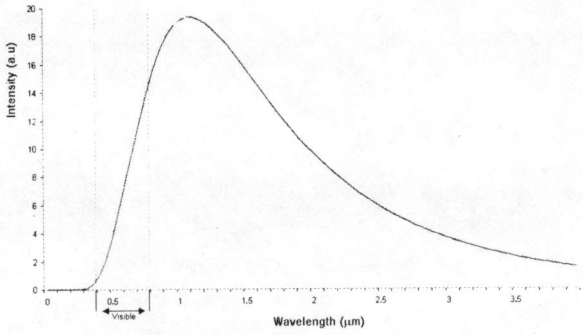

Figure 2.6: *The optical spectrum of a tungsten filament lamp*

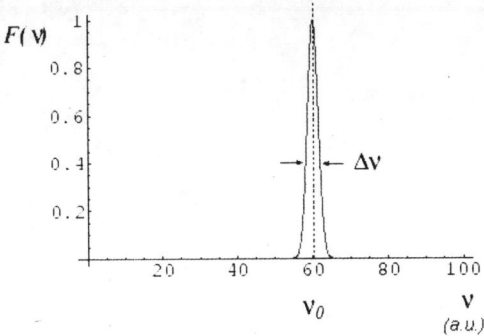

Figure 2.7: *Spectrum of a quasi-monochromatic source*

2.11. Conclusion

Since free-space telecommunication relies on the use of electromagnetic waves, it seemed essential to us to briefly describe some properties of these waves. Likewise, since geometrical optics is extensively employed for simulating free-space links, it was necessary to present the relation between electromagnetic waves and light rays, as well as the practical limitations of geometrical optics.

Of course, the source is of vital importance in an optical link. Therefore, the following chapter will turn to the description of the sources used for wireless telecommunication.

Chapter 3

Emission and Reception of Optical Beams

3.1. Foreword

The light sources and receivers which we consider in this chapter are hybrids of semiconductor components. Most of them are manufactured industrially and used in multimode and single mode optical fiber links: they are already manufactured and available commercially at low cost in great quantities.

This chapter is built on the differences between:

– the essential ratings and characteristics of transmitters and receivers optimized for digital transmissions at high data rates along single mode optical fibers over long distances,

– the essential characteristics of the semiconductor transmitters and receivers dedicated to digital transmission in free-space optical links.

The first characteristic common to both light emitter and detector is the optical spectral range.

In the two spectral windows centered on 0.8 and 1.55 µm wavelengths:

– specific attenuations of multimode and single mode fibers are small,

– manufacturing technologies, materials and single mode fiber structures allow excellent dispersion and amplification performances to be attained,

– passive components – couplers, filters and photo-incorporated Bragg network reflectors – are already industrial realities,

– erbium-doped optical fiber amplifiers (Er^{3+}), high power lasers for optical pumping at 1.48 µm, are available.

And furthermore:

– in ophthalmology, the components of this infrared spectral window are considered less dangerous to high power densities than those centered on shorter wavelengths such as 0.85 µm. They are primarily absorbed by the cornea and do not seem to cause irreversible pathological damage to human eye constituents,

– all the sources, components, couplers and filters quoted previously and used in optical fiber links are available industrially.

3.2. Introduction

In the case of free-space propagation of light waves, requirements are basically different from those for waves guided over long lengths of optical fibers. In free-space optics, the light-wave spatial–temporal mode specific to the transmitter propagates through the air in the direction of a receiving surface some distance away from the emitting element surface. The notion of spatial mode of the fiber is to some extent replaced by the concept of a path which connects the transmitting point on the surface of the source to a point on the surface of the detector.

In this chapter we describe active optoelectronic components, sources and photodetectors; their structures; and their optical and electronic properties. We also mention the electronic circuits which are associated with them and which constitute the emission modules and the photoreception modules.

Some of the active elements of emission and reception functions can be retained among those conceived and used industrially in fiber optic links. In this case, it is necessary to associate them with passive optics elements and operate them under the best conditions of radiometry adapted to their optical characteristics and to the transmission conditions in the atmosphere.

This chapter is dedicated to transmitting and receiving components and comprises three sections:
    – a review of the basic concepts of radiometry /photometry,
    – the transmitting and amplifying components: spectral and spatial properties,
    – "low signals and low noise" photoreceivers.

The optimization of transmission and reception data processing – adaptability of the active components (transmitter and detector) to the electronic circuits, technologies of compensation of the obstacles to the propagation, etc. – will only be mentioned briefly.

3.3. Radiometry: basic concepts

Optical ray paths are the trajectories followed by the wave in the atmosphere, the wave originating in an element in the source, from the pupil − spatial output mode of the transmitter − and propagating toward an element in the receiver area. When the propagation medium is homogeneous, these paths are straight.

The source emits well-defined waves through its pupil output − or electromagnetic modes:
− in wave optics, with their optical frequencies and their radiant power distributions on the pupil area − near field:

Near field $\Psi(x,y,z,t) = \psi(x,y).\exp i(\omega t - kz)$

Optical frequency: $\nu = \dfrac{\omega}{2\pi}$ [3.1]

Radiant power *distribution* : proportional to the pupil area $|\psi(x,y)|^2$

where x and y are the co-ordinates of the elementary source in the pupil area. A source can consist of several source elements of surfaces dS which transmit waves of different spectra according to all the light intensity modes − or radiant power − distributed on the pupil output area. In the most general case represented in Figure 3.1, elementary sources dS_1, dS_2 and dS_3 transmit their own waves along their own axis, their own radiation pattern and their own wavelength spectrum:

− in radiometry (Figure 3.2) with their radiation patterns and their brightness:

$$d\Omega = \frac{dS'.\cos\varphi}{r^2} = \frac{d\Sigma}{r^2}$$ [3.2]

$$I(\theta) = \frac{d\phi}{d\Omega}; (example : I(\theta) = I_0.\cos(\theta) : Lambert's\ law)$$ [3.3]

$$L(\theta) = \frac{dI(\theta)}{dS.\cos\theta}$$ [3.4]

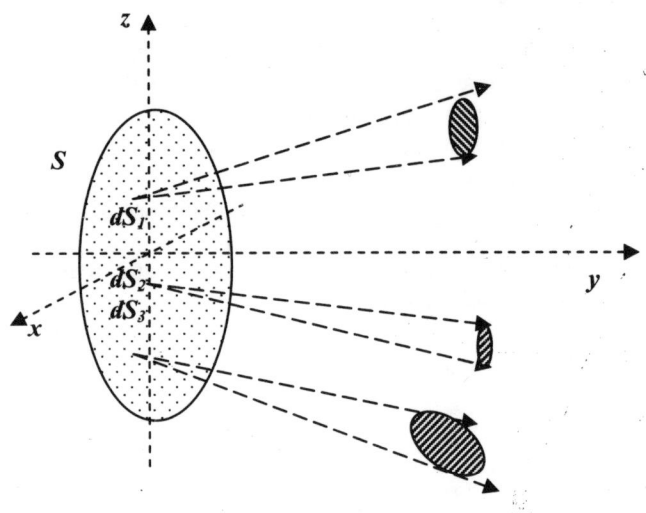

Figure 3.1: *Concept of emitting area*

Each point O, center of a surface element dS on the emitting area, transmits its waves inside a cone where O is the top and $d\Omega$ is the solid angle defined by the output pupil of surface dS'. This surface element is characterized by the normal vector which forms an angle ϕ and by the distance r which separates it from O. In Figure 3.2a, the surface element $d\Sigma'$ represents the projection of dS' in the plane perpendicular to the average direction of propagation. The elementary quantity of radiant power $d\Phi$ contained in this cone is proportional to the solid angle element $d\Omega$ itself proportional to $d\Sigma'$. The ratio between $d\Phi$ and $d\Omega$ is the energetic intensity I. This energetic intensity is defined in an elementary cone and can depend on the angle θ, the direction of the solid angle element axis OM in space.

The emitting surface S of a large area light source consists of a set of elementary sources with surface dS which transmit their elementary energetic intensities dI in the direction θ. Energy brightness $L(\theta)$ of the source S is the ratio between the elementary intensity $dI(\theta)$ transmitted in the direction θ and the apparent surface element $d\Sigma$ in the direction θ as represented in Figure 3.2 (b).

The radiant power $d\Phi$ of a surface element dS is expressed in W/m², the total radiant power which propagates through the output pupil is expressed in W, the energetic intensity $I(\Theta)$ in W/sr and the energetic brightness in W/sr.m².

From these relations, definitions and given the orientation and the brightness distributions of the source surface, we can calculate the total radiant power available in the plan of an output pupil and the illumination E_{tot} – the radiant power – at the surface of the detector received by this as represented in Figure 3.2(c) [SURREL, 2000]:

$$d\Omega = \frac{dS'.\cos\theta'}{r^2}$$

$$\frac{d^2\Phi}{dS'} = L.\frac{dS.\cos\theta.\cos\theta'}{r^2} \qquad [3.5]$$

$$E_{tot} = \iint L.\frac{\cos\theta.\cos\theta'}{r^2}.dS$$

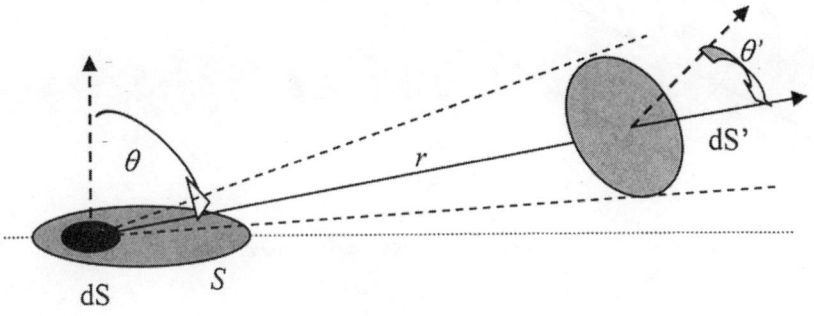

Figure 3.2: *Elements of radiometry: emitting point (dS);
element of a large area light source S; photoreceiver dS': orientation θ'*

For free-space optical links, whatever the application is, the detector receives the maximum radiant power from the light source when the image of this source is formed on its surface inside its zone of maximum sensitivity. Consequently, we can obtain optimum conjugation by using an optical device whose structure and materials depend on the characteristics of the optical field close to the source and the geometrical characteristics of the detector sensitive area. It is the extended-beam concept (Figure 3.3) which leads to the definition of this device. This ensures the development of the source image on the detector surface and the best transport of the energy flux and which closely connects the source to the detector at the same time.

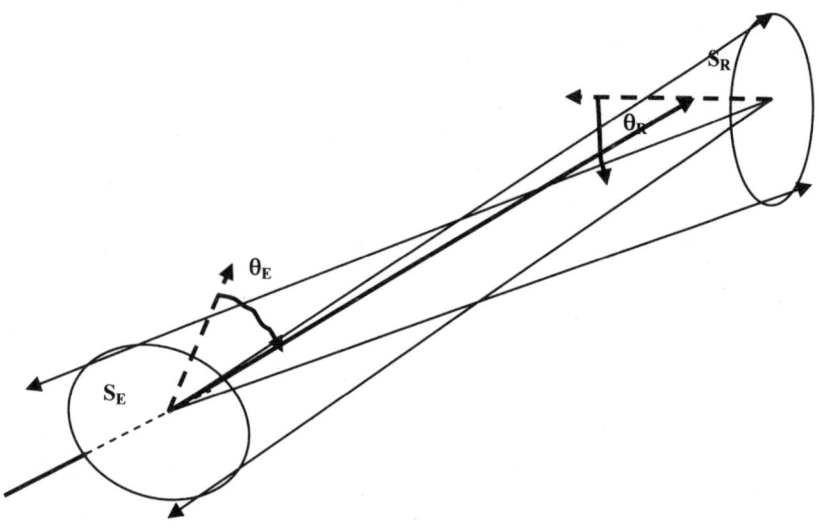

Figure 3.3: *Extended-beam concept*

The characteristics of systems designed for adapting between emitters and receivers according to the propagation conditions are defined from these concepts. Among these devices the most common components are thin plates, lenses and mirrors, which can be associated with more or less complex diffractive components.

Returning to the active elements, the optical and electronic characteristics of sources and detectors depend directly on the materials from which they are made and on their structure. Materials and structures are generally defined from the choice of the data rate, the required transmission quality.

The transmitting and receiving devices can thus be structurally very different from those which are "fiber-pigtailed" and currently used as transmitters and receivers at both ends of single-mode fiber links. These differences are considered and studied in the following two paragraphs.

3.4. Optical spectral windows, materials and eye-safety

Because of its absorption coefficient, silicon can cover the visible −1.0μm range to the border of the 1.0−1.6 μm range. The latter can be covered by other materials which have been developed more recently. Figure 3.4 represents the absorption coefficients α (cm^{-1}) of most common semiconductor materials plotted against wavelength (μm).

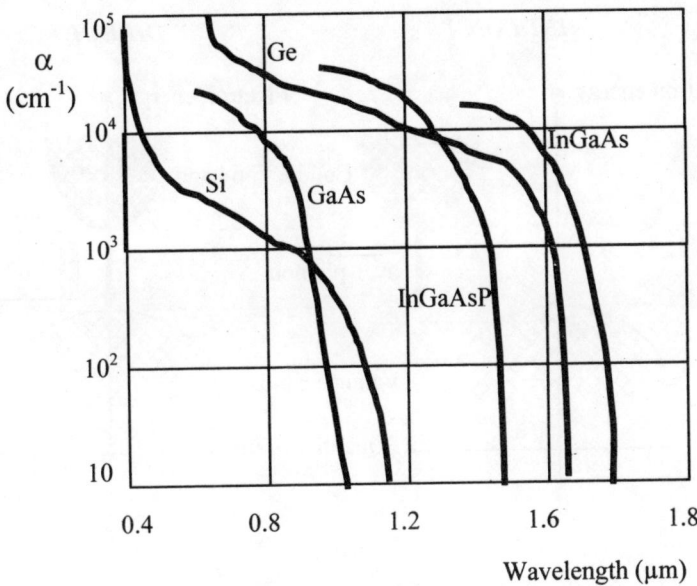

Figure 3.4: *Absorption coefficients of common semiconductors (cm^{-1}) against wavelength (μm)*

Multimode, single-mode, silica and polymer fibers show the best characteristics of propagation in the two spectral windows [0.6 to 1.00 μm] and [0.98 to 1.6 μm]. The light sources of optical waves detected by the silicon detectors are commercially available (for example: opto-couplers, visible display LEDs, LEDs for polymer fiber transmission systems). Light sources and detectors of waves whose wavelengths lie between 0.98 and 1.6 μm have also been developed essentially for fiber optics applications and are also commercially available.

Electroluminescence is the fundamental mechanism of generation of the photons transmitted by semiconductor light sources. They are made from binary, ternary or quaternary materials, elements of groups III and V in Mendeleev's periodic table (aluminum Al, arsenic As, gallium Ga, phosphorus P, indium In). Thin layers of these binary, ternary and quaternary compounds are deposited on substrate wafers - binary compounds GaAs or InP. Their conduction and valence band structures are such that the probability of direct "band to band" transition of carrier electrons and holes is very high. Compared to the "band to band" transitions in materials such as silicon and germanium (indirect bands), for example, these transitions do not require crystal vibrations like phonons and are known as "direct transitions" (see Figure 3.5).

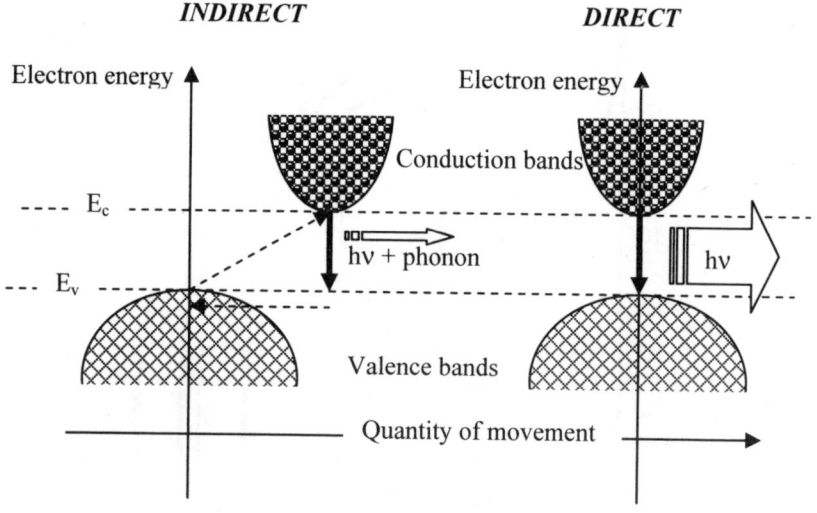

Figure 3.5: *Semiconductors: energy band structures and radiative transitions* *[JOINDOT, 1996]*

Particular attention needs to be given to the materials and active components operating in the spectral window which are least likely to endanger to human sight and human eye constituents, or cause pathological disorders.

The eye-safe spectral window is narrow and lies between 1450 and 1550 nm (Figure 3.6). The first ocular constituent exposed to a light beam is the cornea (Figure 3.7).

Considering the waves belonging to this spectral range, the radiant power is not totally absorbed by the cornea (75%); the rest propagates and is mainly absorbed in the aqueous humor. The radiant power density per unit area transmitted to the aqueous humor is low, so that little damage is done to the crystalline lens which floats in this medium, and a fortiori to the retina.

In addition, the cornea has a resistance to light exposure equivalent to the resistance of the skin, as well as a very high capacity for regeneration. Most of the radiant power is thus far absorbed by the most "protective" constituent of the human eye. The requirements for ocular safety are published and available from several international standardization organizations such as the International Electrotechnical Commission (IEC) [PRUNNOT, 2002].

Some spectral components of the rare earth erbium ion Er^{3+} in silica and glass matrices, corresponding to transitions between energy level and resonances, are located around 1.55 µm. For this reason, 1.55 µm is the central wavelength of Er^{3+} doped fiber amplifiers and at the origin of the industrial development of low loss − low dispersion fibers and optical components for Dense Wavelength Division Multiplexing (DWDM) transmission systems.

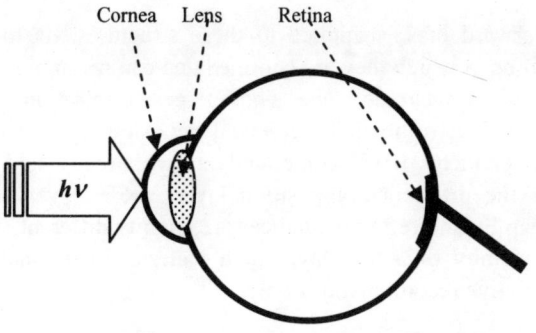

Figure 3.6: *Schematic cross-section of an eye*

3.5. Transmitters

The properties of thermal radiation sources do not correspond to what we expect and what we think is needed and useful from light transmitters in terms of brightness, radiation characteristic and spectrum. Most of the light sources which are used are quantum devices, and most commonly lasers. Lasers are the brightest sources, among them: solid lasers (YAG-Nd or ruby); gas lasers (HeNe, metal vapors excited by electronic discharge); lasers with amplifying optical fibers doped with rare earth ions and pumped optically; and semiconductor electroluminescent lasers.

Note (about the modulation in information transmission) – the radiant powers – or light intensities – of gas lasers and most of those which are pumped optically (such as YAG-Nd) cannot be directly – i.e. internally – modulated. External modulators must then be used and inserted in the beam to act on the intensity or on the phase of the wave.

For these reasons, electroluminescent diodes and semiconductor junction lasers have a great advantage since their radiant powers and optical frequencies can be directly modulated by the injection current. Modulation frequencies up to several tens of GHz can be reached; these performances depend on the material and the internal structure of the components.

3.5.1. *Broad spectrum incoherent light emitting diodes*

3.5.1.1. *Structures*

When a forward bias is applied to these structures, the injected carriers are driven into a zone in which they are confined and can recombine radiatively inside a planar low loss waveguide: one photon per electron-hole pair. There are homojunctions or heterojunctions, epitaxial depositions and superposition of thin layers of semiconductor III–V compounds as we already said above. The lattice parameters of the different composition layers are not very different and their energy band gap E_G and refractive indices are slightly different. The recombinations occur in the vicinity of a thin layer of a material intentionally left undoped to support the radiative recombinations only.

3.5.1.2. *Near and far field patterns*

The structures of two light emitting junctions are represented in Figures 3.7a and 3.7b. In the case of emission through the top layer surface, we usually consider that the propagation of photons inside the radiative region is isotropic. The photons which are able to escape from the device propagate inside a cone whose half angle is limited by the refraction conditions at the diopter level between the crystal and the ambient. The others are reflected back into the radiative transition region. In this case, the radiation pattern is a sphere tangential to the top surface (Figure 3.8a). The near field pattern depends on the upper contact geometry and the carrier injection conditions inside and around this contact area.

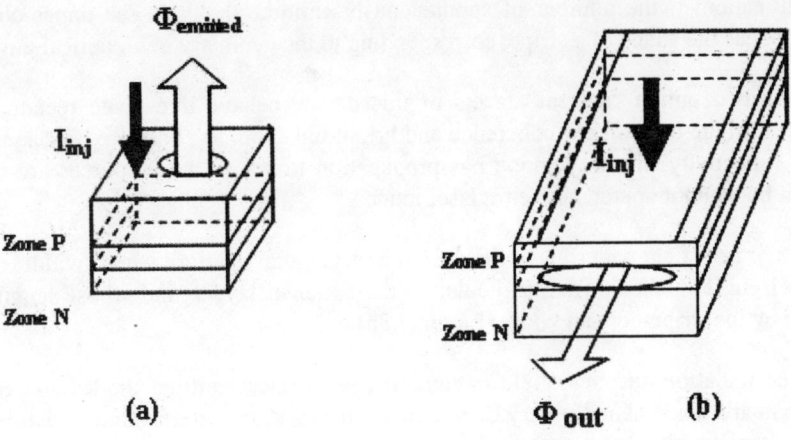

Figure 3.7: *Examples of electroluminescent diode sources of incoherent light waves:*
(a) surface emission, (b) emission through the edge facet (edge-emitter)

The internal quantum efficiency (IQE) is the ratio of the number of photons created in the device to the number of injected electron-hole pairs. The external quantum efficiency is the IQE divided by the reflection coefficient at the interface between the material and ambient. Performance in terms of internal and external quantum efficiency can be optimized by adding epitaxial layers to confine both the carriers and the photons, by modifying the reflection coefficient of the anti-reflecting layers deposited on the upper active layer surface, etc.

The near field pattern initially imposed by the aperture of the contact (Figure 3.7a) and the radiation pattern of the structure (Figure 3.7b) can be modified using suitable lenses and/or passive optical micro-systems. When we improve the directivity of a light source we have to increase the size of the near field pattern.

When the radiative recombination zone is localized inside a heterojunction, the internal quantum efficiency can be very close to unity. The compositions of the adjacent zones, and consequently their energy gaps and refractive indices, are optimized to confine the maximum number of carriers and photons in a loss-less planar structure parallel to the layers. This structure, known as an edge emitting diode, can then guide the photons towards the facets of the "chip" obtained by cleavage along natural crystalline parallel planes. It is also considered as very "bright" or "super-luminescent". Under strong injection of carriers, there is inversion of the populations between the two sets of levels of carriers and simulated emission of photons. It can therefore operate in stimulated emission mode with amplification of the number of spontaneously emitted photons. The upper ohmic contact has the shape of a stripe, corresponding to the geometry of an optical guide.

The two output "natural" facets of this device behave like plane rectangular mirrors. In the case of low coherence and bright light sources, both are processed to have reflectivity; the waveguide has propagation losses so these sources are very bright but cannot operate and emit laser modes.

Their near field pattern is defined from a rectangular aperture whose width is the optical guide thickness, perpendicular to the epitaxial layers, and whose length is given by the stripe-contact width (Figure 3.8b).

The radiation and near field patterns of top surface emitting diodes and edge emitters are fundamentally very different. Their radiation patterns, and variation of light intensity with the direction of propagation, can be represented by one family of analytical functions:

$$I(\theta) = I(0).(\cos\theta)^n = I_0.(\cos\theta)^n \qquad\qquad [3.6]$$

in which the directivity increases with the exponential n.

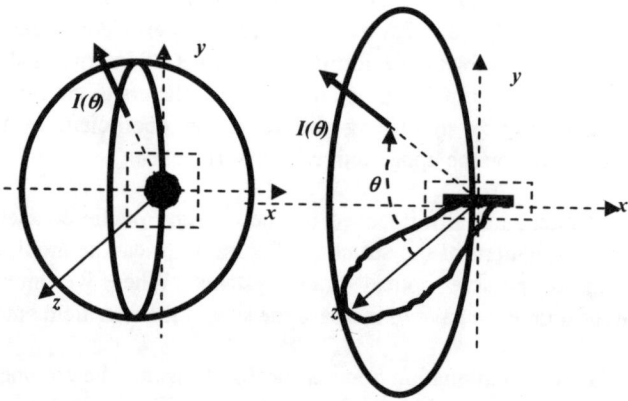

(a) Lambert's law: $I(\theta) = I(0).\cos\theta$
spherical revolution around Oz

(b) $I(\theta) = I(0).(\cos\theta)^n$
n(xOz) different from n(yOz)

Figure 3.8: *Radiation patterns and near field patterns of electroluminescent diodes*

In the case of edge emitter diodes (Figure 3.7b), the output facet is a section of a rectangular dielectric waveguide. The distribution of the electric field vector module is Gaussian along the two axes of the facet plan. The modulus squared is proportional to the radiant power per unit area. Therefore, the near field pattern, the distribution of the radiant power density on the output facet, has a Gaussian shape. The far field, which is the spatial Fourier transform of the near field, also has a Gaussian shape the top of which is centered on the axis of the waveguide. The double integral of the following expression is carried out on the surface of the facet/pupil; u and v are the approximation of a current point coordinates at the surface of the far field:

$$\left|\Psi(u,v)\right|^2 \approx \frac{1}{\lambda * R^2} * \left|\iint \psi(x,y) * \exp\left[-2i\pi(u*x+v*y)\right]dx*dy\right|^2 \qquad [3.7]$$

with $u = \dfrac{\sin\theta_x}{\lambda}$ and $v = \dfrac{\sin\theta_y}{\lambda}$.

$\psi(x,y)$ is a gaussian in x and $y \Rightarrow \Psi(u,v)$ is a gaussian in u and v.

3.5.1.3. Spectral characteristics

A typical spectrum of these sources is represented on Figure 3.9. The spectral width $\Delta\lambda$ is usually represented by the difference in wavelength between the components at half height of the maximum amplitude component λ_p and varies with the temperature according to the following approximate relation [TOFFANO, 2001]:

$$\Delta\lambda(\mu m) = 1.45 * \lambda_p^2 * [k_B T] \quad [k_B T] \text{ is expressed in } (eV) \qquad [3.8]$$

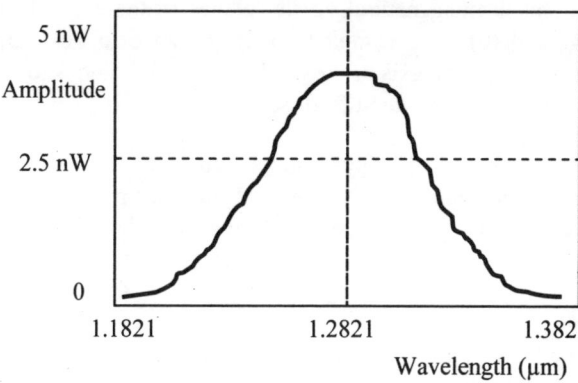

Figure 3.9: *Spectrum of the incoherent light wave emitted by an electroluminescent diode*

These broad spectrum components are considered as incoherent wave sources compared to lasers, with coherent lengths not exceeding a few microns.

Note – these transmitters are particularly well adapted to transmissions along index gradient multimode fibers and are commercially available in fiber pigtailed packages with several tens of microns diameter circular near fields.

3.5.1.4. Electrical and optical characteristics

The average values of electrical and optical essential characteristics of these optical broadband sources at room temperature are given in Table 3-1.

	Surface emitting diode	Super luminescent diode
Output radiant power (I_F = 100 mA)	1 mW	2 to 3 mW
Optical response time to a current step	from 3 to 5 ns	about 1 ns
Electrical power	200 mW	200 mW

Table 3-1: *Standard characteristics of optical broadband transmitters*

3.5.2. Laser diodes: high radiant power output, coherent waves

3.5.2.1. Structures

We now consider the structures of super luminescent edge emitting diodes. They are planar dielectric optical waveguides with rectangular section along which the carriers recombine, generating photons; and inside which the created photons can propagate and be amplified. These guides lie inside resonant cavities between two parallel facets oriented along natural crystal planes in the case of "Fabry–Pérot" type lasers (Figure 3.10). The resonant structure can also be a Bragg grating, periodic corrugation along the axis of the guide, whose function is to select only one single mode in the case of distributed feedback lasers (DFBs).

When the injection current I is high enough and exceeds the so-called threshold current, the gain is practically equal to the total losses of the cavity, which are the propagation losses of the guide (such as scattering, absorption, defects at the interfaces between thin epitaxial layers) and the losses at the reflective facets of the resonator.

In general, the active part of a semiconductor laser is the core of the dielectric waveguide buried in thin layers of materials of various compositions and electronic and optical properties. The reader will find a detailed description of all the technologies of manufacture of the semiconductors lasers in the book *Technologies pour les composants à semi-conducteurs – principes physiques* (Technologies for semiconductors components – physical principles) [FAVENNEC, 1996].

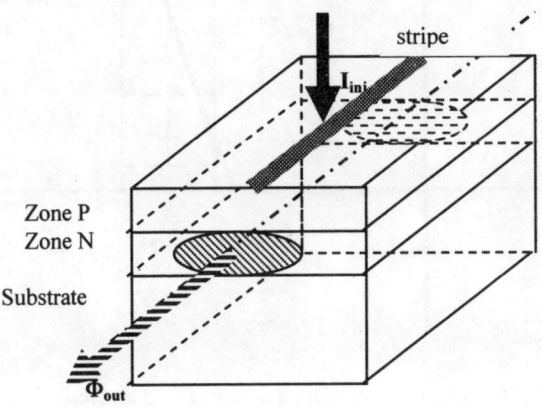

Figure 3.10: *Laser structure: thin epitaxial layers, stripe and Fabry–Pérot resonator*

3.5.2.2. "($\Phi_{transmitted}$)/($I_{injected}$) characteristic": static and dynamic

A typical curve is represented in Figure 3.11. The resonator is a Fabry–Pérot cavity laser emitting wavelengths around 1.33 µm. Some typical values of radiant power output and direct modulation performances are given in Table 3.2 below.

	Single transverse mode stripe-geometry lasers (F-P and DFB)	High power lasers
DC radiant power output	5 to 10 mW	> 500 mW
Max. direct intensity modulation frequency	> 20 GHz	To be specified

Table 3-2: *Some typical values of radiant power output and direct modulation performances*

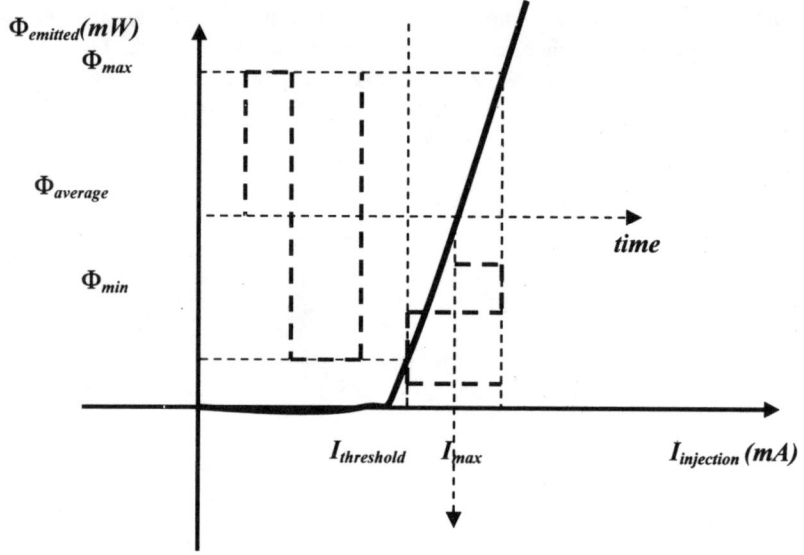

Figure 3.11: *DC radiant power plotted against $I_{forward}$ and modulation response*

3.5.2.3. *Spectra and near field patterns*

These "Fabry–Pérot" structures can transmit several waves distributed in one or more modes of transverse and/or longitudinal wave forms. In Figure 3.12, we show the result of the spectral analysis of the near field of a 1.3 μm laser. The spectrum and near fields were scanned at a forward injection current close to threshold.

The entire source spectrum shines through a Czerny–Turner spectrometer on an infrared vidicon-type camera. We can simultaneously observe:

– in the upper part of the Figure: four patterns. Each pattern consists of four emission areas which show the presence of transverse modes from 1st to 4th orders;

– in the lower part of the Figure: the spectral envelope of the four longitudinal modes whose emission areas have a pattern aligned on the horizontal axis. The envelope is sampled.

In the case of fiber links or in integrated optics, the 0-order transverse modes whose bright areas have one spot-pattern only along the laser axis can be coupled and guided into a single mode fiber (or single mode waveguide). All the other modes may then be re-considered in free space optical wave propagation.

Thus far, the lasers optimized for transmission applications along single mode optical fibers are those whose:

– spectra correspond to the best conditions of propagation and to the minimal temporal dispersion,

– near-field pattern geometries and radiation patterns are matched to the core and the acceptance angle of the fiber to obtain the best coupling efficiencies.

Their spectra contain:

– longitudinal modes in the case of the Fabry–Pérot type lasers whose facets are cleaved according to two parallel crystalline plans (Figure 3.11): spectra from Figure 3.13 (on the left and in the centre) corresponding respectively to the "Fabry–Pérot" lasers with gain and index guidance,

– single mode imposed by the Bragg network engraved along the active guide in the case of lasers known as "DFB on distributed negative feedback": Figure 3.13 (on the right) (DFB: Distributed Feedback Lasers).

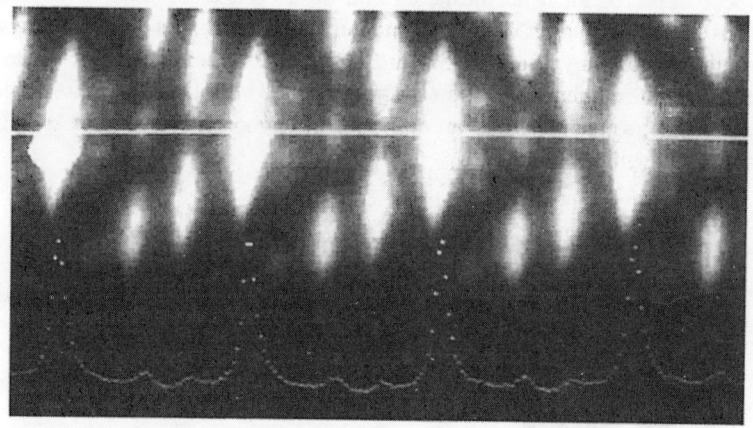

CYB 2002-2003

Figure 3.12: *Near field (top) and spectrum (bottom) of a semiconductor laser close to threshold: multimode emission on several transverse modes*

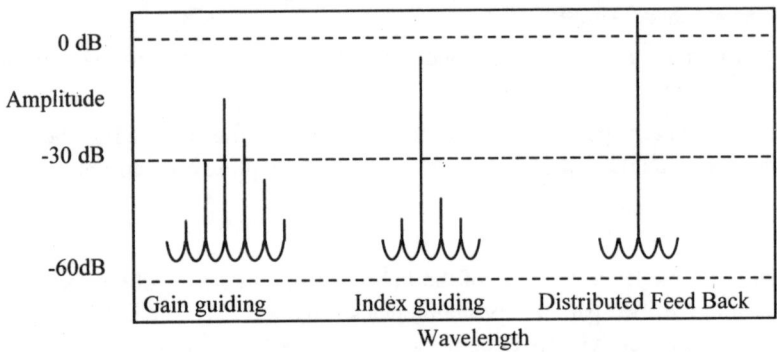

Figure 3.13: *Spectra of semiconductor lasers. The amplitude of each component is referenced to the total radiant power output; the ratios are expressed in dB along the vertical axis*

New high power semiconductor laser structures have been industrially developed. Some of them are designed and optimized to be used in optical fiber amplifiers as optical pumps. They are quantum well structures whose energy band configurations and alloy compositions are represented schematically in Figure 3.14. The threshold current densities of these structures are very low. The waveguides and the facets of the resonant cavities are designed and processed to support high radiant power densities without failure.

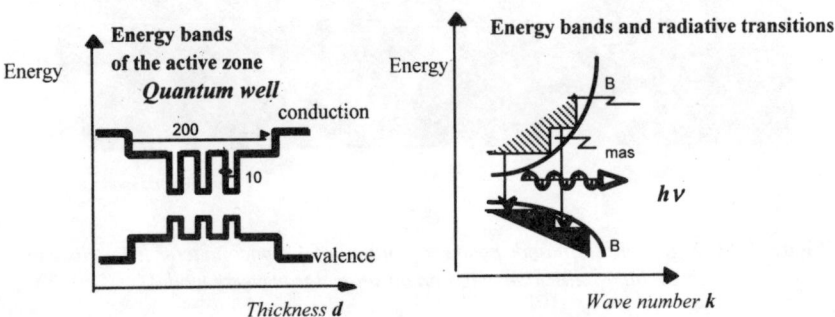

Figure 3.14: *Quantum well structures: details of epitaxial layers and energy band configuration (to be compared to Figure 3.5) [TOFFANO, 2001]*

High radiant power laser-chips are now industrially available, optimized in spectrum, energy flux, reliability and lifetime to be used as "optical pumps" in rare earth ion-doped fiber optical amplifiers in the wavelength spectral range [0.98 to 1.48 µm]. For example, the erbium-doped fiber amplifier spectral range is centered on $\lambda = 1.55$ µm wavelength. Their characteristics are very different from those of the preceding lasers for the following reasons:

– these devices are designed and processed to work CW at normal electrical and thermal rates and to deliver very high radiant powers without particular spectral and radiometric characteristic requirements apart from their stability in time,

– their characteristics in direct and fast modulation are not considered among the reference points of operation and essential characteristics by the standardization institutions (CEI),

– in the case of the transmitter modules assembled with the most recent passive optical coupling elements, all (or most of) the modes delivered by the laser chip are coupled to a "fiber pigtail" designed and positioned to obtain the best coupling efficiency of laser modes into the amplifying fiber. This short-length fiber element (<1m) is then joined by fusion to the amplifying fiber in such a way that the modes which are used for optical pumping include the mode which carries the information and needs to be amplified. Among the best coupling solutions which lead to the highest and most efficient interaction between the pump wave and the propagating mode to be amplified along the amplifying fiber, we find associations of high radiant power output laser chips to multimode fibers. Some examples of near field patterns obtained by simulation [LEPROUX, 2001] are represented in Figure 3.15.

The near field and radiation patterns of the waves transmitted by these high radiant power modules are thus very well defined from the output faces of the "pigtail fiber" elements. The "ideal" uniform emission surface is obtained when the pigtail fiber structure allows the mixture and chaotic interference of all modes. However, it remains to be confirmed that:

– all the modes available from the "pigtail fiber" have the same direct modulation frequency response with the injection current,

– the competition between modes of different order are sources of instability of light intensity or sources of noise intensity which would have to be added to the shot and thermal noise sources of the receiver.

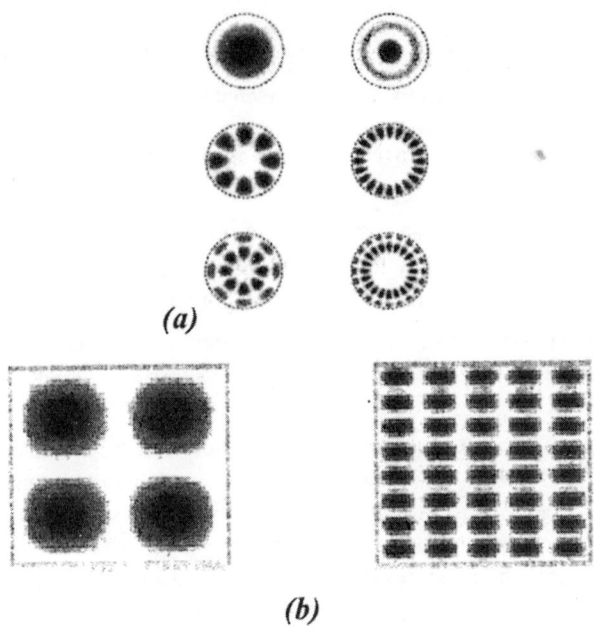

(a)

(b)

Figure 3.15: *Multimode "pigtail fibers" coupled to high power optical pump lasers. Near field patterns of spatial modes. (a) Cylindrical fiber, (b) Rectangular fiber*

3.5.2.4. *Spectral and modal instabilities and light intensity noise*

These phenomena were analyzed in Fabry–Pérot semiconductor laser sources transmitting several longitudinal modes with the same transverse order. The radiant power random fluctuations are also called intensity noise. The curves of Figures 3.16a and b represent the evolutions respectively of:

– the ratio of the noise intensity to the average value of the incident radiant power on the detector, against the injection current referenced to the threshold current ($I/I_{th} - 1$):

$$\text{Noise of relative intensity: RIN} = \frac{\langle \Delta \Phi^2 \rangle}{\langle \Phi \rangle^2} \quad \text{expressed in (Hz)}^{-1} \qquad [3.9]$$

– longitudinal mode distributions of a single transverse mode laser operating between 1.32 and 1.34 μm. If we consider the operating conditions represented by (D), we can see that a chaotic and "conflictual" mode distribution corresponds to a peak in the relative intensity noise, compared to the situation represented by (E) where each longitudinal mode of the cavity exists and the intensity of the whole population remains stable.

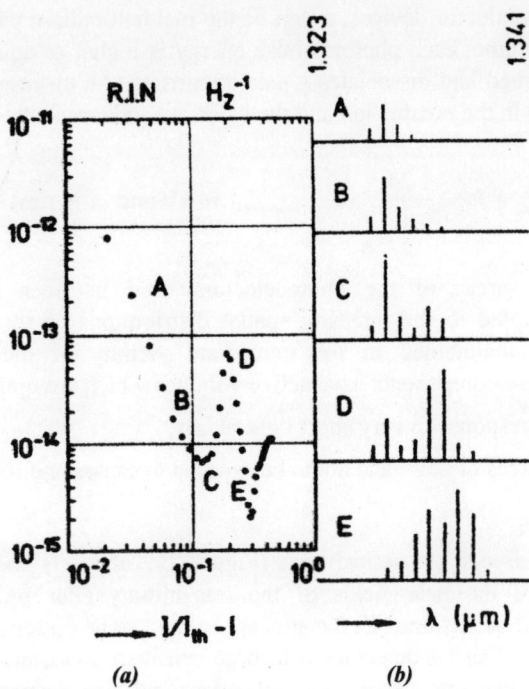

Figure 3.16: *Index guided semiconductor lasers: (a) Relative Intensity Noise against injection current referenced to the threshold current; (b) Longitudinal modes distribution at different injection currents against wavelength*

3.5.3. *Use of amplifiers with "rare earth ion" doped fibers*

Another solution to produce a powerful, low noise and well defined transmitter in spectrum and near field, at the same time, is in coupling a single mode laser transmitter to a doped fiber amplifier. All the elements necessary to assemble such equipment are industrially available in the 1.55 μm wavelength range. Radiant powers of the order of one watt can be obtained from a single transverse mode light source of 10 μm in diameter and the acceptance angle of the "regular telecom" fibers.

3.6. Photodetectors

Photodetectors, particularly those made of semiconductors, are either photo-resistances, or current or voltage generators. When they are illuminated, the values of the resistance, the current or the voltage depend on the incident radiant power. In the case of semiconductor devices, a part of the incident radiant power is absorbed in the detector volume: each photon whose energy is higher or equal to the energy band gap is absorbed and dissociates a pair of carriers. An electron and a hole are therefore released in the conduction and the valence bands respectively:

$$E_{photon\ absorbed} = h\nu \geq \frac{1.24}{\lambda} \quad (E_{photon\ absorbed}\ \text{in eV and}\ \lambda\ \text{in}\ \mu\text{m}) \quad\quad [3.10]$$

The sensitive areas of the photodetectors used in fiber optic links are geometrically adapted to the intensity spatial distributions of the guided modes. These areas are maintained in the immediate vicinity of the fiber endface. Consequently, these components have active volumes which are optimized:

– to be fast in response to very short light pulses,

– not to be sources of noise and not to be exposed to causes and sources of ambient noise.

In the case of free-space optical links (Figures 3.2 and 3.3), the photodetectors receive images of the near fields of the transmitters after propagation along trajectories, optical beams through the atmosphere and some optics elements. A first approach consists of using detectors with both sensitive areas and apertures large enough to capture these images whatever the sizes and displacements of the beams under conditions of maximum illumination. A second approach, technically more sophisticated, consists of using the photodetectors adapted to optic fiber links and fixing them in the image plane of an auto-focusing optical system. Under these conditions the incident field remains inside the perimeter of the detector sensitive area and the incident radiant power is close to maximum.

3.6.1. *Optical spectral range and materials*

As we mentioned above about absorption/sensitivity at the beginning of this chapter:

– silicon is the most appropriate semiconductor material developed for the visible and near infrared wavelengths (450 nm $< \lambda <$ 1000 nm),

– germanium has been replaced by the ternary compound $In_{0.47}Ga_{0.53}$ deposited in thin layers on a binary compound substrate InP (indium phosphide) for the near infrared wavelengths (900 nm $< \lambda <$ 1700 nm),

– the variation of the absorption coefficient vs wavelength depends on the band structure of the semiconductor (Figures 3.4 and 3.5).

3.6.2. *Principle of operation and structures*

In free-space optical links, the incident optical wave illuminates a plane boundary between two different media:

– the propagation medium, the air, whose refractive index is close to 1,

– the semiconductor material whose refractive index lies between 3.4 and 3.7 at the considered wavelengths.

The transmitted part of the incident wave propagates inside the semiconductor which absorbs the photons. Electron-and-hole pairs are then dissociated and released as it has been described above. These two "photo" carriers contribute to the flow of the photocurrent, by opposition with the free carriers which contribute to the dark current. These parasitic free carriers are present in the semiconductor, thermally released or generated by defects of crystalline structure, traps and unstable surface conditions.

3.6.2.1. *Surface phenomena: optical reflection, charge mobility and current leakage*

3.6.2.1.1. Reflection

As mentioned above, the refractive indices of the propagation medium and the illuminated material are very different: approximately equal to 1 for the ambient air and 3.5 for a semiconductor. A part of incident wave radiant power is thus reflected at the surface of the detector when this surface is not covered by "anti-reflecting" dielectric layers. The radiant power which is reflected depends on the refractive indices, on the incident wave angular conditions and on the nature and quality of the antireflective dielectric layers. Without antireflecting coating and under incidence normal to the top surface, the reflection ratio – reflected radiant power to incident radiant power – is given by the Fresnel formula:

$$R(\lambda) = \left(\frac{n_2 - n_1}{n_2 + n_1} \right)^2 \qquad\qquad [3.11]$$

where n_1 and n_2 are the indices of the propagation medium and of the semiconductor material respectively. Without "antireflecting" layers, the "air-semiconductor" interface reflects approximately 30% of the incident radiant power.

3.6.2.1.2. Surface charges and surface currents

The leakage currents whose frequency spectral densities are more important in the low frequency range and whose intensities are proportional to the "active" area of the detector are caused by fixed and mobile electric charges, structural defects and traps present at the interface "ambient-semiconductor material". They are expressed in terms of noise power spectral density at low frequency or "$1/f$ noise". Silicon is a well known material retained for its surface properties, whose oxides are also known for their good stability and technological treatment facilities (passivation).

The photon absorbing region should be an insulating pure intrinsic semiconductor with no free "natural" carriers, "buried" in the volume of the detector and/or protected by passivation layers to avoid pollution from the ambient and parasitic leakage currents. A so-called "dark current" flows under no illumination due to:

– pairs of carriers moving across the unprotected surfaces of this region; they might not follow the same "electric" paths as the photocurrent and they are at the origin of the noise spectral density in $1/f$,

– pairs of thermally dissociated carriers which follow the same current lines as the photocurrent and contribute to the dark current component whose frequency spectral density is as broad as the photocurrent spectral density.

3.6.2.2. *Absorption and conduction: semiconductor junctions*

The most efficient semiconductor structures optimized for low radiant power signal detection are junctions:

– P^+ - N^- (or P^- - N^+): thin abrupt junctions,

– or P^+ - I (I for intrinsic) - N^+ homojunctions in the case of silicon, heterostructures and/or heterojunctions in the case of $In_{0.47}Ga_{0.53}As/InP$.

They all have a low intrinsic defect density absorbing region located either at the junction or between two highly doped regions N and P where the carrier lifetimes are very small. In the case of heterostructures, the absorbing region should not contain free carriers and consists of layers whose compositions and thicknesses are defined so that all the spectral components of the incident wave can be absorbed. A simplified configuration of the energy bands is represented in Figure 3.17: it shows clearly that photons must be absorbed there to dissociate the pairs of "photo generated" carriers. The bottom of the conduction band starts being filled with free "photo-generated" electrons; the top of the valence band being simultaneously filled with free "photo-generated" holes.

The intrinsic absorbing region can be considered as an insulator. When an electric field is applied, this helps separate and speed up the free carriers out of it and out of the space-charge region, these free carriers will then recombine at the borders only where their lifetimes are very short. The intrinsic absorbing region must then be located inside a high electric field zone.

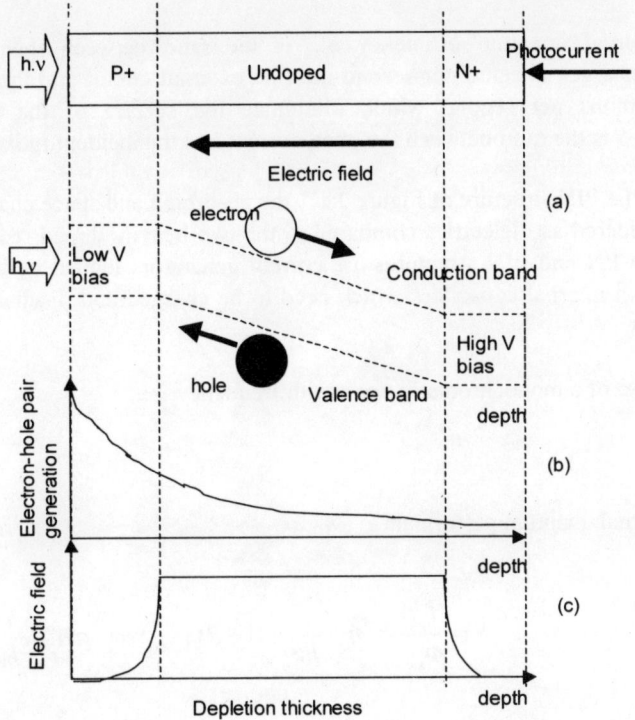

Figure 3.17: *PIN structure: (a) three zones; (b) generation; (c) distribution of electric field*

The electrons are carried away towards zone N and the holes towards zone P. When two carriers recombine at the electrode level, an elementary electric charge is launched into the external circuit. The number of elementary charges delivered per second by the diode under illumination:

$$I_{ph} = N_{e/s} = \eta_{int}.N_{absorbed\ photons/s} = \eta_{ext}.N_{incident\ photons/s} = \eta_{ext}.\frac{\Phi_{inc}}{hv} \qquad [3.12]$$

In other words, the photocurrent is proportional to the incident radiant power. Most of these components can then be considered as photo-current generators loaded by the resistance and the parasitic elements of the external circuit.

3.6.3. *Responsivity, response time, junction capacity and dark current*

The external quantum efficiency η_{ext} is the ratio between the number of elementary charges injected per second into the external circuit and the number of incident photons per second which illuminate the surface of the diode. The responsivity S is the ratio between the photocurrent and the incident radiant power.

Back to the PIN structure of Figure 3.17, the absorbing and space charge regions can be considered as dielectrics compared to the two heavily doped recombination regions. The PN and PIN structures are current generators loaded by a resistance and their own internal capacitors which need to be charged and discharged by the photocurrent.

In the case of a monochromatic wave with frequency ν:

External quantum performance
$$\eta_{ext} = \frac{\dfrac{I_{ph}}{e}}{\dfrac{\Phi_{inc}}{h\nu}} \qquad\qquad [3.13]$$

Sensitivity
$$S = \frac{I_{ph}}{\Phi_{inc}} = \eta_{ext} \cdot \frac{e}{h\nu} \approx (1-R).[1-\exp(-\alpha W)].\frac{e}{h\nu}$$

where e is the electron charge, $h\nu$ the value of the incident photon energy, α the absorption coefficient and W the thickness of the absorbing region under full depletion.

For such a structure, the response to a light pulse depends on two parameters:

– the transit time of the photo-carriers, defined as the time it takes to get to the recombination region through the space charge zone; this time is related to the mobility of the carriers which depends on the applied electric field,

– the time constant of the external circuit load. This time constant is the product of the total capacitance of the current generator in parallel and the load resistor of the external circuit, RC.

We thus reveal the first parasitic element: the total capacitance of the photo-detection first stage, sum of the internal capacitance of the junction and the input capacitance of the external circuit. In practice, the junction capacitance is the major parasitic element in the case of large area detectors such as those which we would advise for free-space optical applications.

The second parasitic element is the dark current. As we have already explained, this current is the sum of several factors:

– the thermal contribution, which depends on the junction temperature and the energy band gap of the semiconductor space charge region,

– the current proportional to the number of transitions per tunnel effect in the case of low energy band gap materials,

– the surface leakage currents which correspond to surface charge mobility and recombination.

In the case of PIN structures, the responsivity and the dark current on the one hand and the capacity of junction on the other hand respectively reach their maximum and minimum values when the electric field completely sweeps the free carriers out of the absorbing and space charge regions. An increase in the applied voltage above this electrical field specific value does not lead to substantial changes in these three essential characteristics.

The power spectral density of the electrical signal developed by the photocurrent in the photo-detector load resistance adds to:

– the spectral densities of the photocurrent shot noise power,

– the light source excess quantum noise power,

– the spectral densities of noise power developed by the dark current and by the leakage currents,

– the thermal noise power density developed by the load resistance.

In practice, for all industrially available photodiodes optimized for optical fiber transmissions, the noise power densities developed by leakage currents at high frequencies are very small compared to the contributions of load resistance, thermal noise and photocurrent shot noise.

Thermal noise, shot noise and light source quantum noise powers are represented analytically by their spectral densities expressed in A^2 /Hz:

$$\langle i^2 \rangle_{thermal} = \frac{4kT}{R_{charge}} \quad (k\text{: Boltzmann's constant})$$

$$\langle i^2 \rangle_{shot} = 2e \langle I_{ph} + I_{dark} \rangle \quad (e = 1.6 \cdot 10^{-19} \text{Coulomb}) \quad [3.14]$$

$$\langle i^2 \rangle_{quantum} = RIN. \langle I_{ph} \rangle^2$$

and, in the equivalent circuit, by current sources which are added to the photocurrent (see Figure 3.18).

From these equations we can highlight the operational limits of PIN structure photo-detectors calculating the optimal photocurrent corresponding to balanced contributions of the three sources of noise.

The reduction of the load resistance noise down to room temperature dark current shot noise leads to very high resistance values, which is in contradiction with the frequency bandwidth/time response of the photo-detector which might be required. Under the conditions of low incident radiant power and large information bandwidth, shot noise power densities are negligible compared to thermal noise power densities of load resistances at the amplifier level.

When the light source is the near field of a multimode source, the light source relative intensity noise (RIN) can exceed the detector shot noise and background noise by several microamperes in photocurrent.

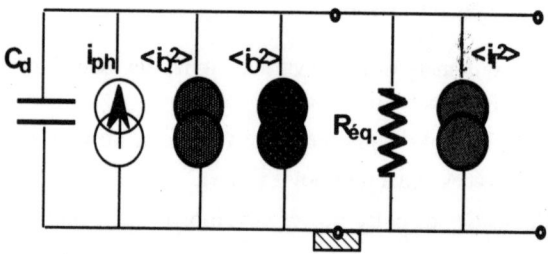

Figure 3.18: *Noise equivalent circuit: photocurrent, parasitic elements and noise sources [TOFFANO, 2001]*

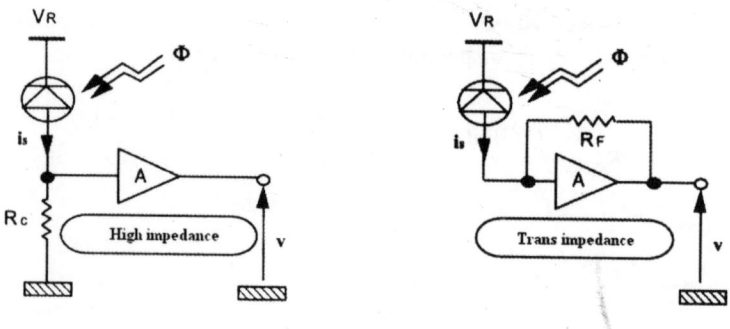

Figure 3.19: *Two examples of circuits [TOFFANO, 2001]*

In the case of links requiring large information bandwidths, we note that the shot noise related to the photons counting process is far too small to appear in the detectable minimum power calculation. What can we expect in free-space optical links from structures with internal gain whose geometries and surface quality are more elaborate and have been successfully optimized for fiber optics links.

3.6.4. Photomultipliers and semiconductor avalanche photodiodes

When the electric field applied across the junction is close to the semiconductor material breakdown limit, the free carriers receive enough energy from the field to collide and hence ionize the surrounding atoms and multiply. Both electrons and holes can multiply under these conditions. The statistics of the multiplication gain in semiconductor structures are more complex than those in vacuum tube photomultipliers. In the latter, electrons are the only charge carriers. Starting from the vacuum tube dynode-to-dynode current relations, the noise properties can be analyzed and derived from the same statistical relations and from the same calculations of averages applied to the crystalline medium.

The noise phenomenon caused by the multiplication of impact and collisions is represented by a factor called Excess Noise Factor (F) in the expression of the current noise spectral density. This factor depends on the ionization coefficients α and β of the two types of carriers which themselves depend on the material in which the ionization takes place. This factor would actually be reduced if only one type of carriers could ionize, which is the case of the vacuum tube photomultipliers. Theory and experiments show that the ideal consists of supporting one of the two types of carriers in "SAGM" heterostructures, with the separation of absorption and multiplication regions, as shown in Figure 3.20:

– in the space charge region, separation of the lower band gap absorption region from the high electric field space charge region, so that only one type of carrier is authorized to cross the multiplication region, get enough energy and ionize,

– a thick multiplication region inside the space charge region to be able to reach a high multiplication gain with very different ionization coefficients under the effect of an optimal electric field,

– a homojunction with a high band gap to limit the dark current components (leakage current and its thermal variations).

Under these conditions, the current noise spectral density increases slowly with the applied voltage and consequently slowly with the multiplication gain M. These variations are represented by the two curves of Figure 3.21:

– the gain M, ratio between the multiplied photocurrent I_{pH} and the primary photocurrent I_{phpr}, I_{pH}/I_{phpr} against the applied tension V,

– the current noise spectral density against the photocurrent I_{pH}, from which we can calculate the excess noise factor F:

$$\langle i^2 \rangle = 2.e.M^2.F(M).I_{primary} \quad \text{with approximation } F(M) = M^{0.2} \quad [3.15]$$

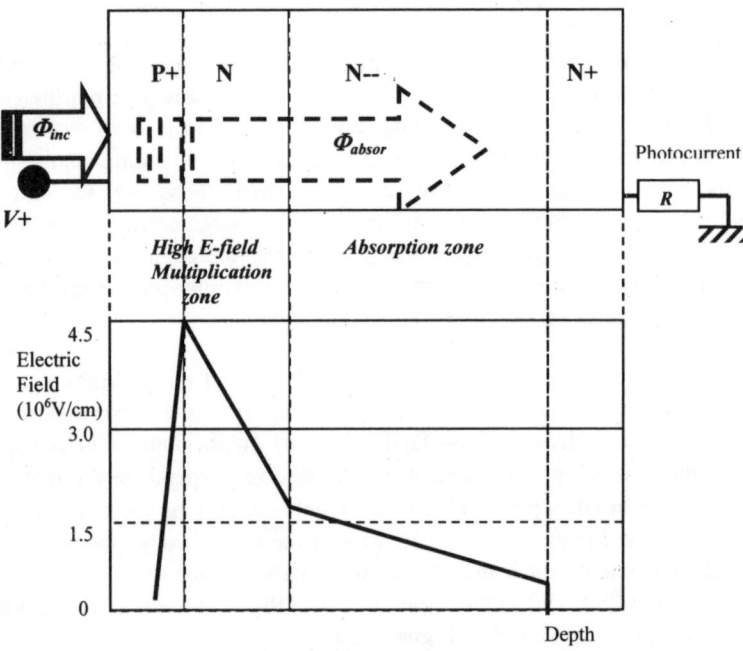

Figure 3.20: *Avalanche photodiode InGaAS/InP with separated absorption and multiplication (SAGM). Top: schematic section, bottom: electric field distribution*

The electric signal-to-noise ratio is given below. In this expression, we assume the light source quantum noise density (RIN) is much smaller than the other sources of noise power densities:

$$\left(\frac{S}{B}\right)_{PDA} = \frac{\left(M \cdot S \cdot \Phi_{incident}\right)^2}{2e \cdot (S \cdot \Phi_{incident} + I_{dark}) \cdot B \cdot M^2 \cdot F(M) + 4kT \cdot B/R_{output.}} \quad [3.16]$$

where:
- e is the electron charge,
- k is the Boltzmann constant,
- T is the ambient temperature in Kelvin,
- $\Phi_{incident}$ is the incident radiant power.

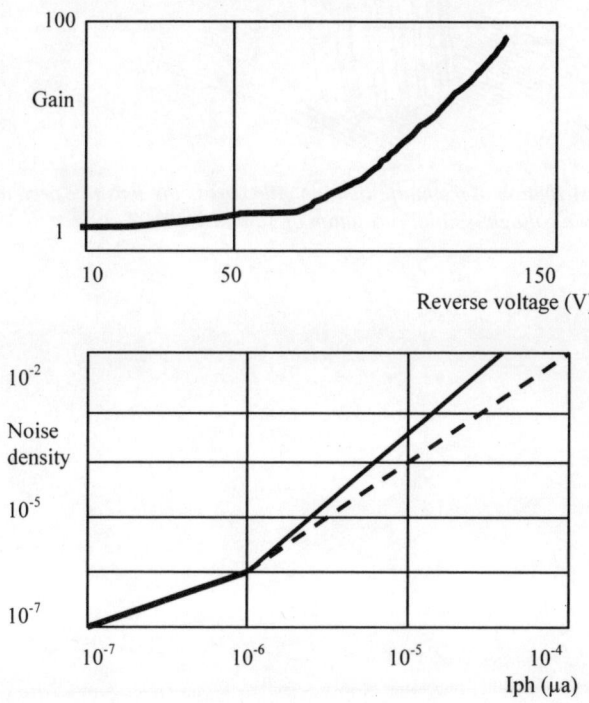

Figure 3.21: *Silicon-SAGM structure: gain and noise characteristics*

We can see the possibility of adjusting the gain acting on the applied voltage so that the magnitude of the first "shot noise" term suitable for photo-detection, together with internal gain and the "second" thermal noise term, are equal. The radio-electric bandwidth and the multiplication factor are represented by the letters B and M respectively; the letter S represents the intrinsic sensitivity of the detector without multiplication gain.

Figure 3.22 represents the amplitude homogeneity of the $M.S$ product. The top surface of a commercially available InGaAs/InP avalanche photodiode is scanned by a 3 μm diameter bright spot (1.55 μm wavelength). The diameter of the active cylindrical volume is 30 μm. This experimental result pointed here highlights one of the technological difficulties one has to solve to industrially develop this type of photo-detector with large homogeneous and stable surfaces.

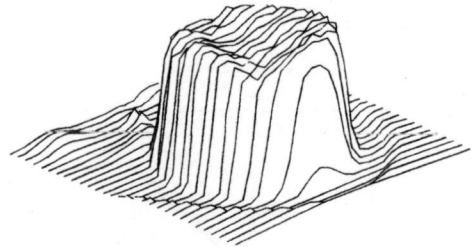

Figure 3.22: *2-D Distribution of the multiplication gain across the sensitive area at 90% breakdown voltage: spatial resolution of 3 μm [JOINDOT, 1996]*

Chapter 4

Line of Sight Propagation

4.1. Influence of the propagation environment

Free-Space Optical (FSO) links involve the transmission, absorption and scattering of light by the Earth's atmosphere. The atmosphere interacts with light due to the composition of the atmosphere which, under normal conditions, consists of a variety of different molecules and small suspended particles called aerosols.

This interaction produces a wide variety of optical phenomena:

– selective attenuation of radiation that propagates in the atmosphere,

– absorption at specific optical wavelengths due to the molecules,

– generation by scattering (the sky blue color, the red sunset, etc.) or by radiative emission of an optical beam comparable to noise at the source of perception contrasts loss. This loss is all the more important since the distance is large,

– scintillation due to the variation of the air's refractive index under the effect of temperature (stars twinkle).

The construction of an optronic system, made up of a transmitter and a receiver in free atmosphere, requires a good knowledge of specific optical properties of the atmosphere like, for example, the spectral transmission of the propagation medium which is affected mainly by the scattering and the absorption of the radiation by this medium. In fact, the performance of any optronic system depends not only on its intrinsic design features resulting from its design and the technology used, but also on its behavior in its operational environment. Thus it is useful to know how optronic systems behave in differing climatic and weather conditions and particularly under prevailing environmental conditions.

To present the effects of the atmosphere on light propagation, it is first necessary to define the following optical phenomena:

– absorption,

– scattering,

– extinction.

4.1.1. *Atmospheric absorption*

Atmospheric absorption results from the interaction between the photons of the radiation and the atoms or molecules of the medium, which leads to:

– the disappearance of the incident photon,

– an increase in the temperature,

– a radiative emission proportional to that of the equivalent black body at the temperature reached.

Let us consider a light beam of wavelength λ which passes through an absorbing medium of thickness dx.

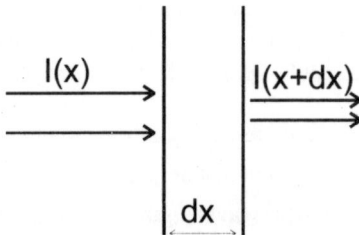

Figure 4.1: *Absorption of light by a thickness dx of absorbing medium*

Due to the absorbent properties of the medium, the number of photons in the radiation is reduced throughout the length of its path. The intensity of the radiation, measured at $x + dx$ (Figure 4.1), in relation to the intensity measured at x, is written as:

$$I(\lambda, x + dx) = I(\lambda, x) - dI_a(\lambda, x) \tag{4.1}$$

The quantity $dI_a(\lambda, x)$ corresponds to the intensity of the light absorbed by the absorbing medium, this latter being proportional to the incident intensity $I(\lambda, x)$, to

dx and to a spectral parameter which represents the absorption of the medium $\alpha(\lambda, x)$ at this wavelength:

$$dI_a(\lambda, x) = \alpha(\lambda, x)I(\lambda, x)dx \qquad [4.2]$$

From these two equations and for a thickness X path, one can write:

$$I(\lambda, X) = I(\lambda, 0)\exp\left[-\int_0^X \alpha(\lambda, x)dx\right] \qquad [4.3]$$

The spectral transmission of the medium is then defined:

$$\tau_a(\lambda, X) = \frac{I(\lambda, X)}{I(\lambda, 0)} = \exp\left[-\int_0^X \alpha(\lambda, x)dx\right] \qquad [4.4]$$

Note – if the propagation medium is homogeneous, the absorption coefficient $\alpha(\lambda, x)$ will be independent of x and the medium spectral transmission is written in the following form:

$$\tau_a(\lambda, X) = \exp\left[-\alpha(\lambda)X\right] \qquad [4.5]$$

4.1.2. Atmospheric scattering

Atmospheric scattering results from the interaction of a part of the light with the atoms and/or the molecules in the propagation medium, which causes an angular redistribution of this part of the radiation with or without modification of the wavelength. To calculate the transmission of a scattering medium we proceed from the preceding paragraph to write:

$$\tau_d(\lambda, X) = \frac{I(\lambda, X)}{I(\lambda, 0)} = \exp\left[-\int_0^X \beta(\lambda, x)dx\right] \qquad [4.6]$$

where $\beta(\lambda, x)$ is the specific spectral scattering coefficient.

Note – if the medium scatters at the same wavelength as the incident radiation, we have Rayleigh and Mie scattering, otherwise we have Raman scattering.

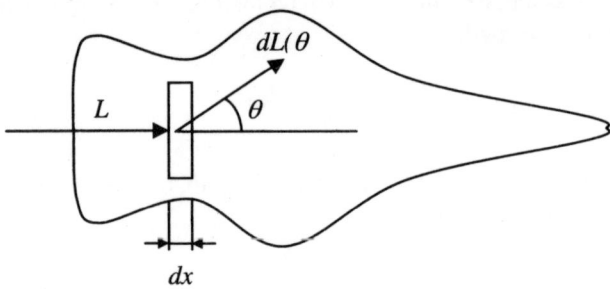

Figure 4.2: *Scattering indicatrix*

In the case of scattering, the scattered light does not disappear locally as with absorption. A scattering medium is characterized by the scattering indicatrix $dL(\theta)$ (Figure 4.2) which describes the spatial distribution of the light scattered per unit of volume. It has symmetry of revolution around the direction of the incident light and it is commonly known as "phase function".

4.1.3. *Extinction and total spectral transmission*

If the propagation medium is both an absorbing and a scattering medium; extinction occurs. The extinction coefficient $\gamma(\lambda, x)$ is defined as the following:

$$\gamma(\lambda, x) = \alpha(\lambda, x) + \beta(\lambda, x) \qquad [4.7]$$

The total spectral transmission relation is then written:

$$\tau(\lambda, X) = \tau_d(\lambda, X)\tau_a(\lambda, X) = \exp\left[-\int_0^x \gamma(\lambda, x)dx\right] \qquad [4.8]$$

4.1.4. *Earth's atmosphere*

For Free-Space Optical links, the propagation medium is the atmosphere. The atmosphere may be regarded as a series of concentric gas layers around the Earth. From 0 to 80–90 km of altitude there is the homosphere; beyond this is the heterosphere. Three principal atmospheric layers are defined in the homosphere: the troposphere, the stratosphere and the mesosphere. These layers are differentiated by their temperature gradient with respect to the altitude.

In Free-Space Optical telecommunication, we are especially interested in the troposphere because this is where most weather phenomena occur. The propagation of light in the troposphere is influenced by:

- the gas composition of the atmosphere,

- the presence of aerosols, that is, small particles of variable size (ranging from 0.1 to 100 µm) in suspension in the air,

- hydrometeors such as rain, snow, hail,

- lithometeors such as dust, smoke, sand,

- modifications of the gradient of the refractive index of the air (propagation medium) at the source of scintillations and turbulence.

4.1.4.1. *Atmospheric composition*

In order to characterize the properties of atmospheric transmission affecting optronic systems, the gas components of the atmosphere are classified in two categories:

- the components with fixed density proportion or majority components (their variation is smaller than 1%), they have a quasi-uniform distribution for altitudes ranging up to between 15 and 20 km. The most important among these constituents are nitrogen (N_2), oxygen (O_2), argon (Ar) and carbon dioxide (CO_2). In the visible and infrared regions, up to a wavelength of 15µm, CO_2 gives the only important absorption lines,

- the components with variable density, they are in the minority and their concentration depends on the geographical location (latitude, altitude), on the environment (continental or maritime) and on the weather conditions.

Water vapor is the main variable constituent of the atmosphere. Its concentration depends on climatic and meteorological parameters. While its concentration may reach 2% in maritime environments, its presence is negligible at altitudes higher than 20 km.

The water content is determined from the atmospheric humidity and can be defined in three different ways:

- the absolute humidity (g/m^3) gives the mass of water vapor per unit air volume,

- the relative humidity (%) can be defined as the ratio between the absolute humidity and the maximum quantity of vapor that could be contained in the air at the same temperature and at the same pressure,

- the number of mm of precipitable water (w_0) per unit distance, usually per km.

Another major variable component is ozone (O_3) whose concentration varies also with altitude (maximum content at 25 km), latitude and season. It presents an important absorption band in ultraviolet, and in the infrared region around the 0.6 μm wavelength.

4.1.4.2. Aerosols

Aerosols are extremely fine particles (solid or liquid) suspended in the atmosphere with a very low fall speed caused by gravity. Their size generally lies between 0.01 μm and 100 μm. Due to the action of terrestrial gravity, the biggest sized particles ($r > 0.2$ μm) are in the vicinity of the ground. Fog and mist are liquid aerosols, salt crystals and sand grains are solid aerosols.

The presence of aerosols may cause severe disturbance to the propagation of optical and infrared waves, since their dimensions are very close to the wavelengths of these frequencies. It is not the same in the range for instance of centimeter and millimeter waves where the wavelength is much longer than the size of the aerosols.

Atmospheric attenuation results from an additive effect of absorption and dispersion of the infrared light by gas molecules and aerosols present in the atmosphere. It is described by Beer's Law giving transmittance as a function of distance:

$$\tau(d) = \frac{P(d)}{P(0)} = e^{-\sigma d} \qquad\qquad [4.9]$$

where:

- $\tau(d)$ is transmittance at the distance d of the transmitter,

- $P(d)$ is the power of the signal at a distance d of the transmitter,

- $P(0)$ is the emitted power,

- σ is the attenuation or the extinction coefficient per unit of length. Attenuation is related to transmittance by the following expression:

$$Aff_{dB}(d) = 10 * \log_{10}\left(1/\tau(d)\right) \qquad\qquad [4.10]$$

The extinction coefficient σ is the sum of four terms:

$$\sigma = \alpha_m + \alpha_n + \beta_m + \beta_n \qquad\qquad [4.11]$$

where:

- α_m is the molecular absorption coefficient (N_2, O_2, H_2, H_2O, CO_2, O_3), etc, please refer to the structure and the composition of the atmosphere),

- α_n is the absorption coefficient of the aerosols (fine solid or liquid particles) present in the atmosphere (ice, dust, smoke, etc.),

– β_m is the Rayleigh scattering coefficient resulting from the interaction of light with particles of size smaller than the wavelength,

– β_n is the Mie scattering coefficient, it appears when the incident particles are of the same order of magnitude as the wavelength of the transmitted wave.

Absorption dominates in the infrared whereas scattering dominates in the visible and the ultraviolet regions.

4.2. Visibility

4.2.1. *Generalities*

4.2.1.1. *Definitions*

Visibility was defined originally for meteorological needs, as a quantity estimated by a human observer. However, this estimation is influenced by many subjective and physical factors. The essential meteorological quantity, namely the transparency of the atmosphere, can be measured objectively and it is called the Runway Visual Range (RVR) or the meteorological optical range.

In the atmosphere, the runway visual range is the distance that a parallel luminous ray beam, emanating from an incandescent lamp, at a color temperature of 2700 K, must cover so that the luminous flux intensity is reduced to 0.05 of its original value. The luminous flux is evaluated using the photometric luminosity function of the "commission internationale de l'éclairage" (CIE). Daytime runway visual range and night time runway visual range are distinguished [OMM, 1989].

The daytime runway visual range is defined as the longest distance at which an appropriately sized black object, located in the vicinity of the ground, can be seen and identified when it is observed on a sky or on a fog scattering background.

The night time runway visual range is defined as the longest distance at which an appropriately sized black object can be seen and identified if general illumination is increased to normal daylight intensity, that is, the longest distance at which moderated intensity light sources can be seen and identified.

The light of the air is the light of the sky and the sun scattered towards the eyes of an observer by suspended particles in the air (and, to a smaller extent, by the air molecules) located in the observer's field of view. The light of the air is the fundamental factor which limits the diurnal horizontal visibility for black objects because its contribution, integrated along the field-of-view between the observer and the object, increases the apparent brightness of a sufficiently distant black object to a level which cannot be distinguished from that of the sky background. At the opposite of a subjective estimate, most of the light of the air which penetrates into the observer's eye finds its origin in portions of its near field-of-view.

The following photometric qualities are defined with reference to standard works such as *50(845)* the CEI standard of the International Electro-technical Commission [CEI50, 1987]:

– luminous flux (symbol F (or Φ); unit: lumen, lm) is derived from radiant energy flux by calculating the radiation according to its action on the CEI standard photometric observer,

– light intensity (symbol: I; unit: candela, cd or $lm.sr^{-1}$) is the luminous flux per unit solid angle,

– brightness (symbol: L; unit: $cd.m^{-2}$) is the light intensity per unit area,

– illumination (luminosity) (symbol: *E; unit*: lux or $lm.m^{-2}$) *is* the luminous intensity per unit area.

The extinction coefficient (symbol a) is the proportion of luminous flux that a beam of parallel luminous rays, emanating from an incandescent source, at a color temperature of 2700 K, loses by crossing a length equal to unit distance in the atmosphere. This coefficient measures the attenuation due to both absorption and scattering.

The brightness contrast (symbol C) is the ratio of the difference between the marker brightness and its back-plan to the brightness of this back-plan.

The contrast threshold (symbol ε) is the minimal value of the brightness contrast that the human eye can detect, e.g. the value which allows an object to be distinguished from the background. The contrast threshold varies according to the observer.

The illumination threshold (E_t) is the minimal illumination value for the human eye to detect point light sources against a given brightness back-plan. The value of the illumination threshold E_t thus varies according to meteorological conditions.

The transmission factor (symbol T) is defined, for a beam of parallel luminous rays, emanating from an incandescent source, at a color temperature of 2700 K, as the fraction of the luminous flux that remains in the beam at the end of a given distance of optical trajectory in the atmosphere. The transmission factor is also called the transmission coefficient. When the path length is well defined, as for a transmissometer, the terms *transmittance* or transmissive capacity of the atmosphere are used. In this case, T is often multiplied by 100 and is expressed as a percentage.

4.2.1.2. *Units and scales*

The meteorological visibility or runway visual range is expressed in meters or kilometers. Its measurement range varies according to application.

If, as for synoptic meteorology, the scale of measured values of the meteorological optical range extends from less than 100 m to more than 70 km, the measurement range could be more restricted for other applications. This is the case for civil aviation in particular, as the upper limit can be 10 km. This range can be reduced still more when it comes to measuring the runway visual range representing the landing and takeoff conditions in reduced visibility. The runway visual range is only required to be between 50 m and 1500 m. For other applications, such as road traffic or sea traffic, the limits can again be different depending on need and location.

Visibility error measurements increase proportionally with visibility, and the measurement scales take account of this factor. This fact is reflected in codes used in synoptic reports through the use of three linear decreasing resolution segments, e.g. from 100 to 5000 m they increase by steps of 100 m, from 6 to 30 km by steps of 1 km and from 35 to 70 km by steps of 5 km. This scale allows one to account for visibility with a resolution higher than the measurement precision, except when the visibility is lower than approximately 900 m.

4.2.1.3. *Meteorology needs*

The concept of visibility is largely used in meteorology in two different ways. Visibility is initially one of the elements that identify the characteristics of a mass of air, especially for purposes of synoptic meteorology and climatology; for which it must be representative of the optical state of the atmosphere. This is then an operational parameter which corresponds to specific criteria corresponding to special applications. It is expressed directly in terms of distance at which it is possible to see markers or predefined lights.

Measurement of visibility used in meteorology should not depend on extra-weather conditions. It must simply be connected to intuitive concepts of visibility and for distances for which the current objects are seen under normal conditions. The meteorological optical range was defined in order to respond to these needs, while adapting to instrumental methods of day time and night time measurement and taking into account relationships with other measurements of visibility.

The meteorological optical range was officially adopted by the OMM as the visibility measurement for general and aeronautical use [OMM, 1990a]. It is also recognized by the International Electro-technical Commission [CEI50, 1987] for Free-Space Optical applications and for visual signs.

The meteorological optical range is related to the intuitive concept of visibility through the contrast threshold. In 1924, Koschmieder, followed by Helmholtz, proposed a value of 0.02 for the contrast threshold ε. Other authors proposed other values. They vary from 0.0077 to 0.06 and even up to 0.2. Under given atmospheric conditions, the lowest value gives the best estimate of the visibility. For aeronautical needs, it is assumed that the contrast threshold ε is higher than 0.02 and a value of 0.05 is taken since, for·a pilot, the contrast of an object (strip marks) compared to the neighboring ground is much lower than that of an object against the horizon. In addition, it is admitted that, when an observer can just see and recognize a black object on the horizon, the apparent contrast of the object is then 0.05. This resulted in the adoption of a transmission factor of 0.05 in the definition of meteorological optical range.

4.2.1.4. *Measurement methods*

Visibility is a complex psychophysical phenomenon, given mainly by the atmospheric extinction coefficient associated with solid and liquid particles in suspension in the atmosphere: this extinction is primarily caused by scattering rather than by the absorption of the light. The estimate of visibility is subject to variations in perception and interpretation in people as well as the characteristics of light sources and the transmission factor. So any visual estimate of visibility is subjective.

Human observations of visibility depend not only on the photometric and dimensional characteristics of the object which is, or should be, perceived, but also on the contrast threshold specific to the observer. At night, it depends on the intensity of the light source, on the back-plan illumination and, if the estimate is made by an observer, on the accommodation of the eye of the observer in adapting to the darkness, and on the illumination threshold specific to the observer. The estimate of night time visibility is particularly problematic. The first definition of night time visibility is given in terms of equivalent diurnal visibility in order to make sure that no artificial modification intervenes in the estimate of the visibility at dawn and at twilight. The second definition has practical applications, in particular for aeronautical needs, but it is not the same as the first and generally gives different results. Obviously, both are vague.

The extinction coefficient can be measured using instrumental methods, and the meteorological optical range can be calculated from this. Visibility is then calculated from the known values of the contrast thresholds and illumination, or by allocating agreed values to these elements. Sheppard [SHEPPARD, 1983] asserts that to hold strictly onto the definition (of meteorological optical range) it would be necessary to assemble a projector and a receiver equipped with appropriate spectral characteristics on two platforms which can be separated, for example along railway tracks, until the factor of transmission is 5%. Any other approach gives only an estimate of the meteorological optical range.

However, fixed instruments are used, on the assumption that the extinction coefficient is independent of the distance. Some instruments measure the attenuation directly and others measure the scattering light to deduce the extinction coefficient.

4.2.2. *Visual estimate of the meteorological optical range*

4.2.2.1. *General*

Visual estimates of the meteorological optical range can be carried out by a meteorological observer using natural or man-built reference markers (a group of trees, rocks, towers, bell-towers, churches, lights, etc.).

In each station, it is advisable to draw up a plan of the reference markers used for the observations by indicating the distance and orientation of each reference marker compared to the observer. This plan must include reference markers appropriate for diurnal and night observations. The observer will also have to give detailed attention to the significant variations of the meteorological optical range with direction.

The observers must have normal vision and be suitably involved. The observations are made without the assistance of optical apparatus (binoculars, telescopes, theodolites, etc.) and not through a window. The observer's eye must be at normal height above ground-level (approximately 1.5 m). Measurements of visibility should thus not be taken from the higher levels of control towers or from any other high building. This is all the more important when the visibility is bad.

When the visibility varies according to the direction of observation, the recorded value can depend on the use for which it is intended. In synoptic messages, the value to be indicated is the lowest value but for reports intended for aviation, it is advisable to follow directives given in publication OMM-no.731 [OMM, 1990].

4.2.2.2. *Estimate of the day time meteorological optical range*

For diurnal observations, visual estimates of visibility give a good approximation to the true value of the meteorological optical range. For these diurnal observations, one must choose, as reference markers, objects located at as great a number of different distances as possible, provided that they are black or almost black and that they are detached on the sky above the horizon. One will thus eliminate, as far as possible, the objects of clear color or close to a terrestrial back-plan. This is particularly important when the sun illuminates the object. A white house is consequently a bad reference marker and a clump of trees a good reference marker, except when it is brilliantly illuminated by the sun. If one is obliged to take, as a marker, an object profiled against a terrestrial back-plan, the distance between the object and the back-plan should be at least equal to half of that which separates the object from the point of observation. A tree located at the edge of a wood, for example, would not be a satisfactory reference marker for the observation of the visibility.

To be representative, observations must be made according to reference markers whose angular dimension is not less than 0.5° at the observer's eye, because a reference marker which subtends a smaller angle becomes invisible at a certain distance, whereas larger reference markers always remain visible under the same conditions (a 7.5 mm diameter hole bored in a cardboard held at arm's length subtends approximately this angle; a reference marker of visibility seen by this hole should thus fill it completely). In addition, a reference marker should not subtend an angle of more than 5°.

4.2.2.3. *Estimate of the night time meteorological optical range*

Methods used to estimate the night time meteorological optical range from visual observations of the perception distance of light sources are described below.

All light sources can be used as visibility markers, provided that their intensity in the direction of the observation is well defined and known. However, it is desirable to use sources considered as point sources and whose light intensity is not stronger in one direction than in another, and is not contained in a too restricted solid angle. Mechanical stability (fixation) and the optical stability of the light source must be controlled carefully.

It is advisable to distinguish the sources known to be isolated, in the vicinity of which there is not any other source or illuminated area, from the grouped sources, even when these are distinct from each other. This affects the visibility of each source considered separately, so only the use of isolated and suitably distributed sources is recommended for measurements of visibility at night.

The observation of luminous reference markers at night can be affected appreciably by the luminous environment, the physiological effects of dazzling and parasitic lights, even those located out of the field of vision, especially when the observation is made through a window. Thus, a true and accurate observation can be carried out only from a dark place, suitably chosen and located outside any room.

In addition, the importance of physiological factors cannot be neglected, since they constitute an important source of dispersion of measurements. It is essential to use only qualified observers, with normal sight. It is also necessary to allow a period of accommodation, from 5 to 15 minutes, to allow to the eye to adapt to the darkness.

In practice, the relation existing between the perception distance of a light source at night and the value of the meteorological optical range can be expressed in two different ways:

– by indicating, for each meteorological visual range value, the value of the intensity of the light source which, placed at a distance from the observer equal to the meteorological optical range, is just perceptible,

– by indicating, for a given intensity source, the correspondence between the perception distance of this source and the meteorological optical range value.

This second way to proceed is easier and more practical because it is not necessary to install variable intensity light sources at various distances. This method thus consists of using existing light or installed sources in the vicinity of the station. The weather services can thus draw up tables giving the meteorological optical range values in relation to background brightness and known intensity light sources. The values to be attributed to the illumination threshold E_t vary considerably according to ambient brightness. It is advisable to use the following values, considered as average values for the observers:

– (a) $10^{-6.0}$ lux at twilight and dawn or when artificial sources provide an appreciable light,

– (b) $10^{-6.7}$ lux during moonlight or when the darkness is not yet complete,

– (c) $10^{-7.5}$ lux in complete darkness or in star light.

	Light intensity (in cd) of just-visible lamps at given RVR distances		
RVR (m)	twilight ($E_t = 10^{-6.0}$)	moonlight ($E_t = 10^{-6.7}$)	darkness ($E_t = 10^{-7.5}$)
100	0.2	0.04	0.006
200	0.8	0.16	0.025
500	5	1	0.16
1 000	20	4	0.63
2 000	80	16	2.5
5 000	500	100	16
10 000	2 000	400	63
20 000	8 000	1 600	253
50 000	50 000	10 000	1 580

Table 4-1: *Relationship between Meteorological Optical Range (RVR) and the intensity of a just-visible point source of light for three values of E_t (threshold of illumination)*

	Perception distance (in meters) of a lamp of 100 candelas		
RVR (m)	Twilight ($E_t = 10^{-6.0}$)	Moonlight ($E_t = 10^{-6.7}$)	Darkness ($E_t = 10^{-7.5}$)
100	250	290	345
200	420	500	605
500	830	1 030	1 270
1 000	1 340	1 720	2 170
2 000	2 090	2 780	3 650
5 000	3 500	5 000	6 970
10 000	4 850	7 400	10 900
20 000	6 260	10 300	16 400
50 000	7 900	14 500	25 900

Table 4-2: *Relationship between meteorological optical range (RVR) and the distance to which a point source of light of 100 candelas is just visible, for three values of E_t. An ordinary incandescent bulb of a power of 100 watts constitutes a light source of approximately 100 candelas*

Tables 4-1 and 4-2 above give the correspondence between the meteorological optical range and the perception distance of the light sources for each method mentioned above and various observation conditions. This system was established in order to help the weather services to choose or install the light sources which would be used for visibility observations at night and, for observers, to write instructions on the calculation of the meteorological optical range (or runway visual range) values.

4.2.2.4. *Estimate of the meteorological optical range in the absence of distant reference markers*

In certain places (aeroplanes, ships, etc.) or because of a restricted horizon (valley or cirque) or in the absence of adapted landmarks, it is impossible to make observations, except for low values of visibility. For such cases, unless it is not possible to carry out instrumental measurements, it is necessary to estimate the values of the meteorological optical range higher than those for which one has markers from the general transparency of the atmosphere. This can be done by noting the clearness with which the most distant landmarks are detached. Clear contours in the relief, with little or no fuzziness in the colors, indicate that the meteorological optical range is higher than the distance between the landmark and the observer. On the other hand, vague or indistinct landmarks indicate the presence of fog or other phenomena which reduce the meteorological optical range.

4.2.3. *Meteorological optical range measurement instruments*

4.2.3.1. *General*

Instruments to measure the meteorological optical range can be classified in two groups:

– those which measure the extinction coefficient or the transmission factor in a horizontal cylindrical tube of air: the attenuation of light is then due to scattering and absorption by the air particles along the light beam path,

– those which measure the scattering coefficient of the light in a small volume of air.

These instruments use a light source and an electronic device including a photoelectrical cell or a photodiode to detect the emitted light beam.

4.2.3.2. *Instruments to measure the extinction coefficient*

4.2.3.2.1. Telephotometric instruments

A number of telephotometers have been developed which measure, during day time, the extinction coefficient and compare the apparent brightness of a distant object with the sky back-plan, for example, Löhle's telephotometer. However, they are not generally used because of their requirement for preliminary direct visual observations. They can nevertheless prove useful for the extrapolation of the meteorological optical range beyond the most distant marker.

4.2.3.2.2. Visual determination instruments by extinction

For night time, a very simple instrument consists of a degraded neutral filter, reducing, in known proportion, the light coming from a distant light source and in such a way that this one is just visible. The reading of this instrument gives a measurement of the air transparency between the light source and the observer, from which one can calculate the extinction coefficient. General precision depends mainly on variations in the sensitivity of the eye and on the fluctuations of the light source flux density. The error increases in direct proportion to meteorological optical range.

The advantage of this instrument is that it allows the calculation of the meteorological optical range for a range between 100 m and 5 km with reasonable precision, by using only three well-spaced light sources.

4.2.3.2.3. Transmissometers

The transmissometric method is most usually used to measure the average extinction coefficient in a horizontal air cylinder placed between a transmitter consisting of a constant and modulated flux light source and a receiver equipped with a photodetector (generally a photodiode located at a parabolic mirror or the fócal point of a lens). The most frequently used light source is a halogen-type lamp or a tube with luminous discharge in xenon. The modulation of these light sources avoids the influence of parasitic solar light. The current from the photodetector determines the transmission factor that allows calculation of the extinction coefficient and the meteorological optical range.

Assuming that the meteorological optical range transmissometric estimates are based on the light loss of a beam of parallel luminous rays, depending on scattering and absorption, they are closely related to the meteorological optical range definition. A good transmissometer, correctly maintained and functioning in its higher precision range, gives a very good approximation to the true meteorological optical range.

There are two types of transmissometer:

– those where the transmitter and receiver are placed in different cases and placed at a known distance from each other (Figure 4.3),

– those where the transmitter and receiver are placed in the same case, the emitted light is reflected by a mirror or back reflector remotely placed (Figure 4.4).

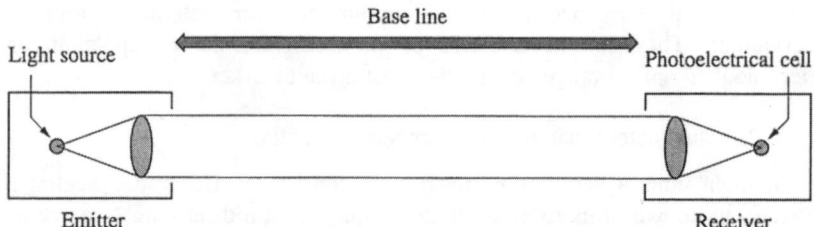

Figure 4.3: *Direct beam transmissometer*

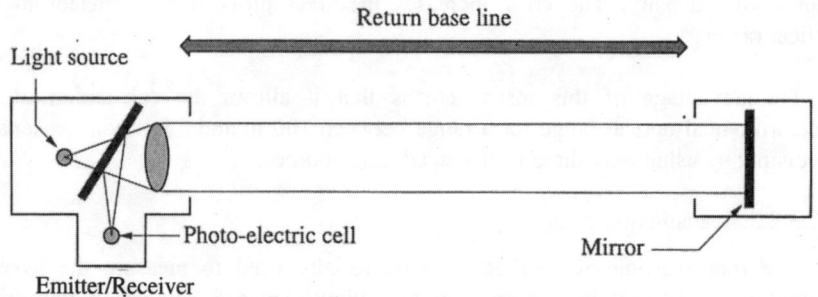

Figure 4.4: *Reflected beam transmissometer*

The distance covered by the light between the transmitter and the receiver is commonly called "transmissometer base" and can vary from a few meters up to 300 m.

In order to ensure measurements remain valid over a long period; it is necessary that luminous flux remains constant during this time. In the case of a halogen light, flux remains constant much longer because halogen lamps only deteriorate slowly and maintain a constant flux for long periods. Nevertheless, some transmissometers use feedback devices (collecting and measuring a small part of emitted flux) which ensure a greater homogeneity of luminous flux in the time or compensate for any modification.

An alternative way of making measurements with a transmissometer consists of using two receivers or back-reflector placed at different distances in order to widen the measured range of the meteorological optical range, both towards the bottom of the range (short base) and towards the top of the range (long base). These instruments are therefore called two-base transmissometers.

For very short base cases (a few meters), one can use a near infrared monochromatic light (electroluminescent diodes). However, it is preferable to use a polychromatic light in the visible range, in order to obtain a representative coefficient of extinction.

4.2.3.3. *Measurement instruments of the scattering coefficient*

Atmospheric light attenuation is due both to scattering and absorption. The absorption factor is important in the presence of pollution in the vicinity of industrial parks, crystals of ice (freezing fog) or dust. However, the absorption factor is usually negligible, and scattering, due to reflection, refraction and diffraction on the water droplets, constitutes the principal factor of visibility reduction. The extinction coefficient can therefore be regarded as equal to the scattering coefficient, and an instrument to measure this scattering coefficient can thus be used to estimate the meteorological optical range.

The most practical method to make this measurement consists of concentrating a light beam on a small volume of air and determining, by photometric means, the proportion of light scattered in a sufficiently large solid angle and not in specific directions. Provided that it is completely protected from other sources of light, such an instrument can be used during both day and night. It is necessary to measure and to integrate the light scattered by the beam for all angles to determine the scattering coefficient precisely. The instruments used in practice measure the light scattered in a limited angle and a high correlation between the limited integral and the complete integral is assumed.

Three types of measurement are used in these instruments: backscatter, forward scatter and scattering integrated on an important angle.

– backscatter (Figure 4.5): the beam of light is concentrated on a small volume of air; it is backscattered and collected by the photoelectric cell.

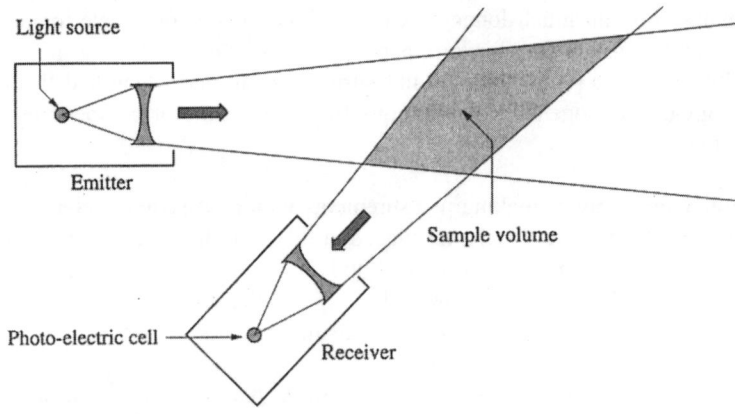

Figure 4.5: *Schematic representation of the measurement of backscatter visibility*

– forward scatter: the instruments consist of a transmitter and a receiver whose emission and reception beams form between them an angle between 20 and 50 degrees (Figure 4.6); other devices place a diaphragm halfway between the transmitter and the receiver, or two diaphragms placed close to the transmitter and the receiver.

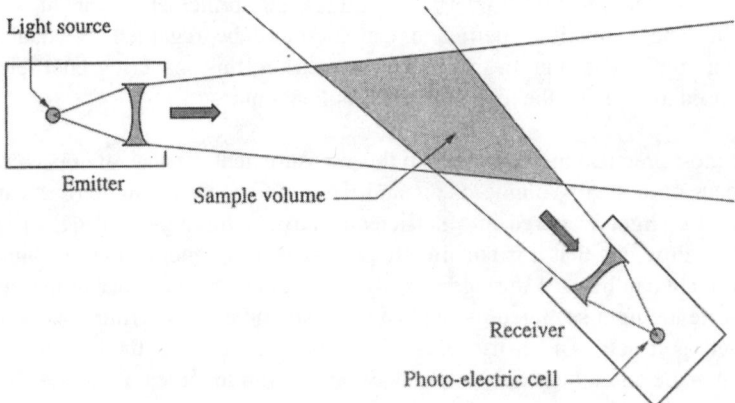

Figure 4.6: *Schematic representation of the measurement of forward scatter visibility*

– scattering under a large angle: the instruments, called integrating nephelometers, use the principle of measuring the scattering for as large an angle as possible, ideally from 0° to 180°, but in practice from 0° to 120°. The receiver is placed perpendicular to the light source axis emitting a light over a large angle. In fact, these integrating nephelometers are not used very often to measure the meteorological optical range but they are frequently used for air pollution measurements.

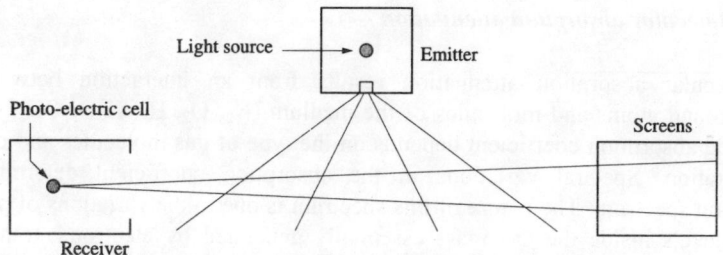

Figure 4.7: *Schematic representation of an integrating nephelometer*

These instruments require only a limited space (generally one to two meters). They are thus very useful when there is no source of light or marker points (ships, roadsides, etc.). For measurements relating to only a very low volume of air, the representativeness compared to the general state of the atmosphere on the site is questionable. However, an average over a number of samples or measurements can improve this representativeness.

The use of this type of instrument is often limited to particular applications (such as the measurement of visibility on motorways or the detection of the presence of fog). The current tendency is to use them increasingly in automatic meteorological observation systems because they allow the measurement of the meteorological optical range over a wide range, and are much less affected by pollution than are transmissometers.

4.2.3.4. *Exposure and implantation of instruments*

The sites of the measuring instruments must be suitably selected. For general synoptic needs, the apparatus should be installed on sites sheltered from local air pollution (smoke, industrial pollution, dusty roads, etc.). The sampled volume of air must be at the observer's eye level i.e. approximately 1.5 m above ground level.

Transmissometers and apparatus to measure the scattering coefficient must be installed in such a way that the sun is not, at any time of the day, in the field of view. Normally, the optical axis is directed horizontally in the North-South direction (+/−45°) or a system of screens or baffles is used for latitudes lower than 50°.

The supports on which transmitters and receivers are used must have good mechanical rigidity in order to avoid any misalignment due to the movement of the ground due to freezing and thawing. The supports should not become deformed due to the thermal stresses to which they are subjected.

4.3. Atmospheric attenuation

4.3.1. *Molecular absorption attenuation*

Molecular absorption attenuation results from an interaction between the radiation and atoms and molecules of the medium (N_2, O_2, H_2, H_2O, CO_2, O_3, Ar, etc.). The absorption coefficient depends on the type of gas molecules and on their concentration. Spectral variations in the absorption coefficient determine the absorption spectrum. The nature of this spectrum is due to the variations of possible energy levels inside the gas mass essentially generated by electronic transitions, vibrations of the atoms and rotations of the molecules. An increase in the pressure or temperature tends to widen the spectral absorption lines by excitation of new possible energy levels and by the Doppler Effect. Molecular absorption is a selective phenomenon which results in the spectral transmission of the atmosphere presenting transparent zones, called atmospheric transmission windows, and opaque zones, called atmospheric blocking windows.

The global transmission windows in the optical range are as following:

- Visible to very near IR   : from 0.4 to 1.4 μm,
- Near IR or IR I   : from 1.4 to 1.9 μm and 1.9 to 2.7 μm,
- Mean IR or IR II   : from 2.7 to 4.3 μm and 4.5 to 5.2 μm.
- Far IR or IR III   : from 8 to 14 μm.
- Extreme IR or IR IV   : from 16 to 28 μm.

The gaseous molecules have quantified energy levels proper to each species. They can, under the influence of incident electromagnetic radiation, absorb energy (or photons) and thus make the transition from an initial level of energy e_i to a higher energy level e_f. The radiation energy is then attenuated by the loss of one or more photons.

This process only appears if the incident wave frequency corresponds exactly to one of the resonant frequencies of the molecule under consideration given by:

$$v_0 = \frac{e_f - e_i}{h} \qquad\qquad [4.12]$$

where:

- v_0 (Hz) is the incident wave frequency,
- h is Planck's constant, $h = 6.6262 \ 10^{-34}$ J.s,
- e_i and e_f are the energy levels of the considered molecular species.

The fundamental parameters needed to determine the absorption generated by molecular resonance are:

- possible energy levels for each molecular species,
- probability of transition from an energy level e_i to an energy level e_f,
- intensity of resonance lines,
- natural profile of each line.

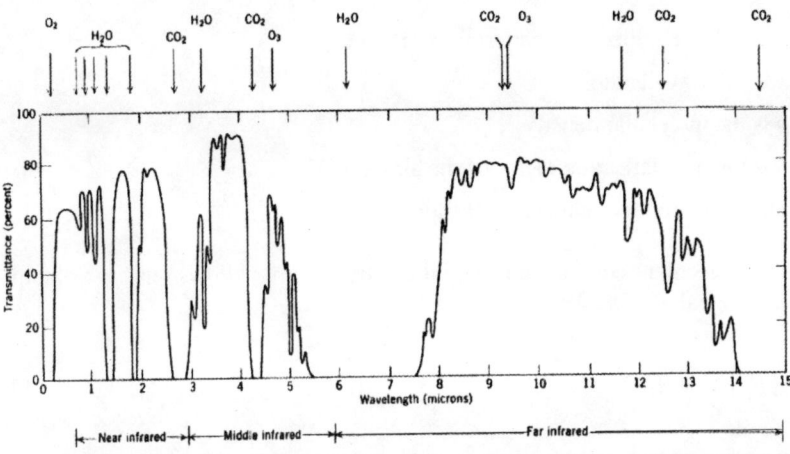

Figure 4.8: *Transmittance of the atmosphere due to molecular absorption*

Generally, the profile of each absorption line is modified by the Doppler Effect when the molecules are moving relative to the incident wave, and by the collision effect due to the interaction of the molecules. These phenomena lead to a spectral widening of the natural line of each molecule. For certain molecules, in particular carbon dioxide (CO_2), water vapor (H_2O), nitrogen (N_2) and oxygen (O_2), the absorption line profiles can extend significantly far from each central line. This property leads to a continuity of absorption with frequency called continuum. Figure 4.8 above gives the variation of the transmittance of the atmosphere resulting from molecular absorption by various components (O_2, H_2O, CO_2, O_3) according to wavelength in the 0.1 to 14 μm range measured on a 1820 m horizontal link at sea level [HUDSON, 1969].

4.3.2. *Molecular scattering attenuation*

The scattering by atmospheric gas molecules (Rayleigh scattering) contributes to the total attenuation of electromagnetic radiation. It results from the interaction of light with particles whose size is smaller than its wavelength.

The expression of the scattering molecular coefficient is:

$$\beta_m(\lambda) = \frac{24\pi^3}{\rho\lambda^4} 10^3 \left(\frac{[n(\lambda)]^2 - 1}{[n(\lambda)]^2 + 2} \right) \left(\frac{6 + 3\delta}{6 - 7\delta} \right)$$

[4.13]

where:

– $\beta_m(\lambda)$ is the molecular scattering coefficient,

– λ is the wavelength,

– ρ is the molecular density,

– δ is the depolarisation factor of the air ($\cong 0.03$)

– $n(\lambda)$ is the refractive index of the air.

The molecular composition of the atmosphere allows us to obtain an approximate value of $\beta_m(\lambda)$:

$$\beta_m(\lambda) = A\lambda^{-4}$$

[4.14]

$$A = 1,09 * 10^{-3} \frac{P}{P_0} \frac{T_0}{T} \ (\text{km}^{-1}\mu\text{m}^4)$$

[4.15]

where:

– P (mbar) is the atmospheric pressure and $P_0 = 1013$ mbar,

– T (K) is the atmospheric temperature and $T_0 = 273.15$ K.

The result is that molecular scattering is negligible in the infrared waveband. Rayleigh scattering is primarily significant in the ultraviolet to visible wave range. The blue color of the clear sky is due to this type of scattering.

4.3.3. *Absorption attenuation by aerosols*

Aerosols are extremely fine solid or liquid particles suspended in the atmosphere with very low fall speed under gravity. Their size generally lies between 10^{-2} and 100 μm. Fog, smoke, dust and maritime spindrift particles are examples of aerosols.

Aerosols influence the conditions of atmospheric attenuation due to their chemical nature, their size and their concentration. In a maritime environment, the aerosols are primarily made up of droplets of water (foam, fog, drizzle, rain), of salt crystals, and various particles of continental origin. The type and density of continental particles depend on the distance traveled and characteristics of neighboring coasts. The absorption coefficient α_n is given by the following equation:

$$\alpha_n(\lambda) = 10^5 \int_0^\infty Q_a\left(\frac{2\pi r}{\lambda}, n''\right) \pi r^2 \frac{dN(r)}{dr} dr \qquad [4.16]$$

where:

 – $\alpha_n(\lambda)$, in km^{-1}, is the aerosol absorption coefficient,

 – λ, in μm, is the wavelength,

 – $dN(r)/dr$, in cm^{-4}, is the particle size distribution per unit of volume,

 – n'' is the imaginary part of the refractive index n of the considered aerosol,

 – r, in cm, is the radius of the particles,

 – $Q_a(2\pi r/\lambda, n'')$ is the absorption cross section of a given type of aerosol.

MIE theory [MIE, 1908] allows us to determine the electromagnetic field diffracted by homogeneous spherical particles and to calculate the absorption Q_a and scattering Q_d cross sections. They depend on the particle size, refractive index, and on the wavelength of the incident radiation. They represent the section of an incident wave: the absorbed (scattered) power is equal to the power crossing this section.

The aerosol refractive index depends on its chemical composition. It is complex and it depends on wavelength. It should be noted that $n = n' + n''$ where is n' is related to the scattering capacity of the particle and n'' to the absorption power of the same particle.

It should be noticed that in the visible and near infrared waveband, the imaginary part of the refractive index is extremely low and can be taken to be negligible in the calculation of global attenuation (extinction). In the far infrared waveband, this is not the case.

4.4. Meteorological disturbances

4.4.1. *Mie scattering attenuation*

4.4.1.1. *Theoretical aspect*

The phenomenon of scattering occurs when the size of particles is of the same order of magnitude as the wavelength of the transmitted wave. In optics this is mainly due to mist and fog. Attenuation is a function both of frequency and of visibility related to the particle size distribution. This phenomenon constitutes the most restrictive factor to the deployment of Free-Space Optical systems over long distances.

Attenuation can reach 300 dB per km whereas rain attenuation in millimeter waves is only about 10 dB per km. The scattering coefficient βn is given by the following equation:

$$\beta_n(\lambda) = 10^5 \int_0^\infty Q_d\left(\frac{2\pi r}{\lambda}, n'\right) \pi r^2 \frac{dN(r)}{dr} dr \qquad [4.17]$$

where:

– $\beta_n(\lambda)$, in km^{-1}, is the scattering coefficient of the aerosol,

– λ, in μm, is the wavelength,

– $dN(r)/dr$, in cm^{-4}, is the particle size distribution per unit of volume,

– n' is the real part of the refractive index n of the considered aerosol,

– r, in cm, is the radius of the particles,

– $Q_d(2\pi r/\lambda, n')$ is the scattering cross section for a given type of aerosol.

Mie theory allows the expression of the scattering coefficient Q_d due to the aerosol. This is calculated on the assumption that the particles are spherical and sufficiently distant from each other so that the field of scattering by a particle and of interaction with another one can be calculated in far field mode (simple scattering).

The scattering cross section Q_d is a function which depends strongly on the size of the aerosol compared to the wavelength. It reaches its maximum value (3.8) for a particle of radius equal to the wavelength: the scattering is then maximal. Then, when the size of the particle increases, Q_d is stabilized around a value equal to 2. This is thus a very selective function which applies to particles of radius shorter or equal to the wavelength. Clearly, scattering depends strongly on the wavelength.

An aerosol's concentration, composition and dimension distribution vary temporally and spatially, so it is difficult to predict attenuation by aerosols.

Although their concentration is closely related to the optical visibility, there is no single particle dimension distribution for a given visibility.

Visibility is a concept defined for meteorological purposes. It is characterized by the transparency of the atmosphere, estimated by a human observer. It is measured according to the meteorological optical range (or runway visual range), the distance that a parallel beam of luminous rays must travel through the atmosphere until its intensity (or luminous flux) drops to 0.05 times its original value. It is measured using a transmissometer or a diffusiometer.

Figure 4.9 gives an example of the variations of the meteorological optical range (or RVR) observed on the site of the TURBIE (France) during one day of high visibility.

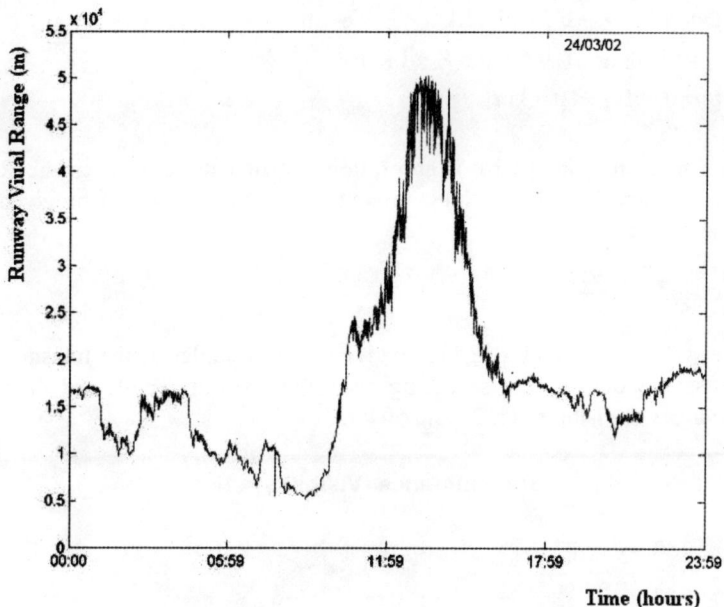

Figure 4.9: *Variation of the Runway Visual Range observed on the site of the TURBIE (France) during one day of high visibility*

The scattering coefficient β_n can be expressed according to visibility and wavelength by the following expression:

$$\beta_n = \frac{3.91}{V}\left(\frac{\lambda_{nm}}{550_{nm}}\right)^{-\varrho}$$

[4.18]

where:
- V is the visibility in km,
- λ_{nm} is the wavelength (in nm),
- Q is a factor which depends on the scattering particle size distribution [KRUSE, 1962]:
  - 1.6 for large visibility ($V > 50$ km),
  - 1.3 for mean visibility ($6 < V < 50$ km),
  - 0.585 $V^{1/3}$ for low visibility ($V < 6$ km).

A recent study [KIM, 2001] proposes another expression for the parameter Q. This expression, not yet proven experimentally, is:

$Q = 1.6$ if $V > 50$ km

$Q = 1.3$ if 6 km $< V < 50$ km

$Q = 0.16 * V + 0.34$ if 1 km $< V < 6$ km

$Q = V - 0.5$ if 0.5 km $< V < 1$ km

$Q = 0$ if $V < 0.5$ km

When molecular and aerosol absorption coefficients as well as the Rayleigh scatter coefficient have low values, the extinction coefficient can be given by the following equation:

$$\sigma \cong \beta_n = \frac{3.91}{V} \left(\frac{\lambda_{nm}}{550_{nm}} \right)^{-Q}$$ [4.19]

Figures 4.10, 4.11 and 4.12 below give some examples of the transmittance of the atmosphere due to Mie scattering according to wavelength and distance for various values of visibility (1, 20 and 50 km).

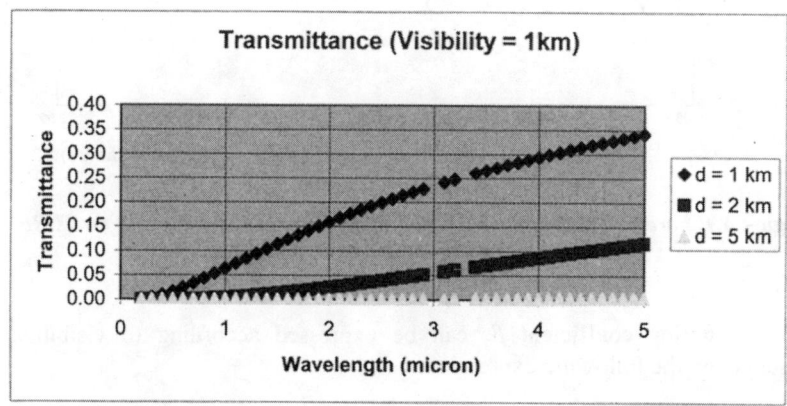

Figure 4.10: *Variation of the transmittance of the atmosphere due to Mie scattering according to wavelength and distance for visibility equal to 1 km*

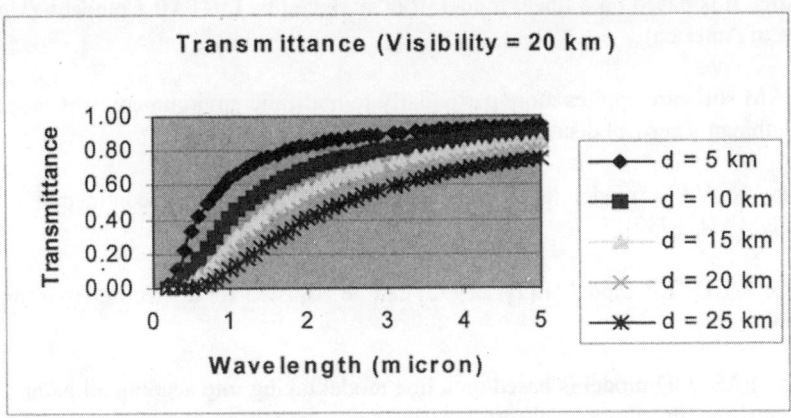

Figure 4.11: *Variation of the transmittance of the atmosphere due to Mie scattering according to wavelength and distance for visibility equal to 20 km*

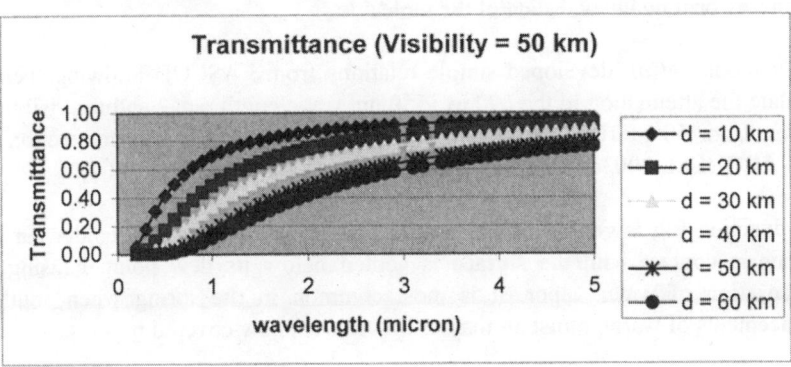

Figure 4.12: *Variation of the transmittance of the atmosphere due to Mie scattering according to wavelength and distance for visibility equal to 50 km*

4.4.1.2. Modeling

A number of computer codes have been developed from the theoretical description of the physical phenomena concerned, to determine the atmospheric transmission coefficient. Several models can be used [COJAN, 1997, DION, 1997, BATAILLE, 1992]: LOWTRAN [KNEIZYS, 1983] and NAM programs (Navy Model Aerosol), NOVAM and WKDAER in maritime environments, NOVAM included in MODTRAN [BERK, 1989], FASTCOD [ONTAR, 1999], etc.

LOWTRAN computer software contains models of optical signal attenuation by aerosols. It is based on a linear model. It is marketed by ONTAR Company (United States of America).

NAM software applies more particularly to maritime environment. It is based on the Gathman's aerosol distribution model [GATHMAN, 1983].

The NOVAM model takes into account dust particles of continental origin [GATHMAN, 1989].

The WKDAER model [LOW, 1992] can be adjusted for a specific environment [DION, 1997].

The FASCOD model is based on a line model taking into account all parameters characterising the lines of absorption (intensity, transition probability, etc.) taken individually (line-by-line calculation). It is based on the high resolution molecular absorption data base HITRAN [ROTHMAN, 1987]. The principal line parameters included in HITRAN are the resonance frequency, the line intensity at 296 K, the probability of transition, the half-width of the line at middle height at 296 K and the low energy or fundamental state of the molecule.

Alnaboulsi *et al.* developed simple relations from FASCOD allowing them to calculate the attenuation in the 690 to 1550 nm wavelength range and for visibilities ranging from 50 to 1000 m for two types of fog: advection and convection fog [ALNABOULSI, 2003a, ALNABOULSI, 2003b, ALNABOULSI, 2004].

Advection fog is generated when warm, moist air flows over a colder surface. The air in contact with the surface is cooled below its dew point, causing the condensation of water vapor. It is most common in the spring when southern displacements of warm, moist air masses move over snow covered regions.

The attenuation by advection fog is expressed by the following equation:

$$\sigma_{advection} = \frac{0.11478\lambda + 3.8367}{V} \qquad [4.20]$$

where:

- λ is the wavelength (μm),
- V is the visibility (km).

Radiation, or convection, fog is generated by the radiative cooling of an air mass during the night when meteorological conditions are favorable (very low velocity winds, high humidity, clear sky). It forms when the surface releases the heat that is accumulated during the day and becomes colder: the air which is in contact with this surface is cooled below the dew point, causing the condensation of water vapor, which results in the formation of a ground level cloud. This type of fog occurs most commonly in valleys.

The attenuation by radiation or convection fog is expressed by the following relation:

$$\sigma_{convection} = \frac{0.18126\lambda^2 + 0.13709\lambda + 3.7502}{V} \qquad [4.21]$$

where:

– λ is the wavelength (μm),

– V is the visibility (km).

There is a polynomial approach to calculate the molecular and aerosol extinction for 6 laser lines (0.83, 1.06, 1.33, 1.54, 3.82 and 10.591 μm), valid at ground level.

These specific wavelengths were selected because they correspond to atmospheric transmission windows for which transmission systems exist (for instance, voice, data, and video images). We describe this model in detail below [BATAILLE, 1992].

4.4.1.2.1. Molecular extinction

The specific extinction coefficient σ_m is obtained by a 10-term expression:

$$\sigma_m = -\ln\left(\begin{array}{l} B_1 + B_2 T' + B_3 H + B_4 T'H + B_5 T'^2 + B_6 H^2 \\ + B_7 T'H^2 + B_8 T'^2 H + B_9 H^3 + B_{10} T'^3 \end{array} \right) \qquad [4.22]$$

where:

– $T' = T(K)/273.15$ is the reduced temperature of the air,

– H, in g/m^3, is the absolute humidity.

The coefficients B_i (i = 1, 10), for the various wavelengths studied, are given in Table 4-3 below.

λ (µm)	0.83	1.06	1.33
B1	0.99528393	0.99300868	1.65651297
B2	0.01311258	0.01959749	−1.92604145
B3	−0.00148875	−0.00227073	−0.00646772
B4	0.00137257	0.00212873	−0.00008351
B5	−0.0121624	−0.0183182	1.88594691
B6	0.00000038	0.00000035	−0.00023356
B7	0.00000026	0.00000067	0.00014466
B8	−0.00042213	−0.0006653	0.00247856
B9	0	0	0.00000045
B10	0.00376331	0.00570955	−0.61624228

λ (µm)	1.54	3.82	10.59
B1	1.00426951	1.31523023	0.46726199
B2	−0.01300905	−0.92958114	1.80965096
B3	−0.00361072	−0.00727529	−0.21839542
B4	0.00342946	0.00867352	0.34811215
B5	0.01322436	0.90633602	−1.6162656
B6	−0.00000698	−0.00011192	−0.00197022
B7	0.00000654	0.00008907	0.00074627
B8	−0.00107642	−0.0030339	−0.13346963
B9	0	0.00000006	0.00001856
B10	−0.00448256	−0.29389689	0.03007365

Table 4-3: *Polynomial formula coefficients of molecular extinction for various spectral lines*

4.4.1.2.2. Aerosol extinction

The specific extinction coefficient σ_n is obtained by a 10-term expression:

$$\sigma_n = -\ln\left(\begin{array}{l} A_1 + A_2H + A_3H^2 + A_4H^x + A_5V^{-1/2} + A_6V^{-y} \\ + A_7HV^{-1/2} + A_8(H/V)^y + A_9H^z/V + A_{10}HV^{-1} \end{array}\right) \qquad [4.23]$$

where:

– V is the visibility in km,

– H, in g/m^3, is the absolute humidity.

– x, y, and z are real numbers used to optimize the polynomial for each studied wavelength. Their values are adjusted when the maximum relative error between FASCOD and the polynomial is lower than 5%.

Coefficients A_i ($i = 1$, 10), for the different studied waves, are given in the tables below for two types of aerosol: rural (Table 4-4) and maritime (Table 4-5).

Figure 4.13 below shows the dependence of aerosol attenuation on wavelength, visibility (for 5 and 25 km) and of environment: rural, urban (which includes soot particles) and maritime (spindrift mainly). It is noted that attenuation decreases with increasing the wavelength. The maritime environment gives the most marked attenuation and it is the rural environment which presents the least attenuation. The curves are for a moderate wind speed and a relative humidity close to 80%. For a relative humidity of 95%, the obtained values would be three times larger [CCIR, 1990].

λ (μm)	0.83	1.06	1.33
x	2.25	4.25	3.25
y	1.188	1.125	1.063
z	5.5	5.75	5.75
A_1	0.134928778	0.079616159	0.04649548
A_2	0.001034237	–0.001638781	–0.001418658
A_3	0.000133549	0.000093129	0.000094803
A_4	0.000042556	–0.000000025	–0.000000748
A_5	0.794668406	–0.424557123	–0.208277287
A_6	1.745654568	–1.314568043	–1.105191514
A_7	0.000514709	0.00021644	–0.000040719
A_8	0.001500683	–0.002042261	0.002653633
A_9	0.000000005	–0.000000003	–0.000000003
A_{10}	0.057904947	0.056277737	0.061228457

λ (µm)	1.54	3.82	10.59
x	5.75	5.25	5.25
y	1.25	2.375	2.375
z	5.75	5.75	5.75
A_1	0.066812858	0.060644611	0.050457822
A_2	−0.00156878	−0.001321056	−0.001434647
A_3	0.000079691	0.000057585	0.000061134
A_4	0	0	−0.000000001
A_5	−0.372419901	−0.311634429	−0.253719348
A_6	−0.648442835	−0.145948587	−0.118015207
A_7	0.000501288	0.001101441	0.001322001
A_8	−0.001018723	0.000012599	0.000012290
A_9	−0.000000002	−0.00000002	−0.000000001
A_{10}	−0.042740204	0.020907523	0.013670880

Table 4-4: *Polynomial formula coefficients of molecular extinction for various spectral lines (rural environment)*

λ (µm)	0.83	1.06	1.33
x	2.75	2.75	3.25
y	1.188	1.188	1.188
z	5.75	5.75	5.75
A_1	0.204552174	0.191893904	0.180960847
A_2	0.000721965	0.002071485	0.002757506
A_3	−0.000051609	−0.000151794	−0.000172694
A_4	0.000002289	0.000007140	0.000001452
A_5	−1.236467317	−1.177820593	−1.123578444
A_6	−2.684606433	−2.545047671	−2.427819252
A_7	−0.000336843	−0.001016813	−0,001286344
A_8	−0.000934423	−0.002401538	−0.002625446
A_9	0.000934423	0.000000002	0.000000001
A_{10}	−0.035349013	0.091877686	−0.099373674

λ (μm)	1.54	3.82	10.59
x	3.75	2.0	5.25
y	1.188	1.125	1.125
z	5.25	2.25	1.50
A_1	0.171528423	0.129576379	0.114747494
A_2	0.003528306	−0.000080764	−0.003090045
A_3	−0.000207060	41.008040428	0.000262312
A_4	0.000000337	−41.008039	−0.000000004
A_5	−1.081749290	−0.734124437	−0.552839212
A_6	−2.339302212	−2.243213505	−1.692154691
A_7	−0.001495886	0.000428726	−0.012032938
A_8	−0.002722432	0.001841746	−0.031923808
A_9	0.000000004	−0.000120611	0.013720388
A_{10}	−0.098225645	−0.001826033	−0.159853116

Table 4-5: *Polynomial formula coefficients of molecular extinction for various spectral lines (maritime environment)*

The formula below [CCIR, 1990] gives an approximate expression for the attenuation due to aerosols according to the visibility (Figure 4.13). It applies more particularly in the visible spectrum (around 0.6 μm):

$$\alpha\left(dB/km\right) = 17/V \qquad\qquad [4.24]$$

where V is the visibility.

Aerosol attenuation (dB/km)		0.98 μm	1.55 μm
Visibility = 25 km	rural model	0.35	0.20
	urban model	0.40	0.26
	maritime model	0.56	0.48
Visibility = 5 km	rural model	1.18	1.02
	urban model	1.22	1.10
	maritime model	1.29	1.26

Table 4-6: *Aerosol attenuation according to the environment, for wavelengths 0.98 and 1.55 μm*

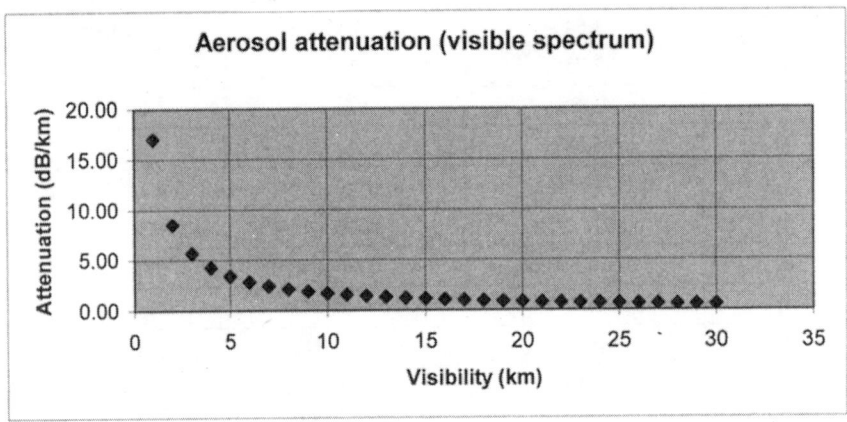

Figure 4.13: *Attenuation in the visible spectrum due to aerosols*

4.4.2. *Cloud attenuation*

When the air cools beyond its saturation point, water vapor condenses to form water droplets, even ice if the temperature is very low. Generally, the water particles thus formed are small in size (<100 μm), but their concentration can be important (a few hundred per cm^3).

The 0°C isotherm defines, from the propagation point of view, the boundary between the liquid and solid phase. The altitude determines the height below which a liquid precipitation (rain) can be observed. The existence of clouds strongly depends on the climate of the region. Cold temperate regions generally have, in summer, minimal cloud coverage and maximal pluviometry. In the Mediterranean region, the opposite is observed. Their presence is characterized by a nebulosity index indicating the fraction of covered sky and expressed in tenths of a percentage point. The multiple scattering effects are important when a light beam travels through clouds. The scattering involves time and frequency dispersion as well as depolarization. Table 4-7 below gives, for certain types of clouds, calculated values of attenuation according to wavelength. Even good weather cumulus causes an attenuation of at least 0.05 dB per meter and nimbostratus (clouds of rain) an attenuation of about 0.5 dB per meter (Figure 4.15).

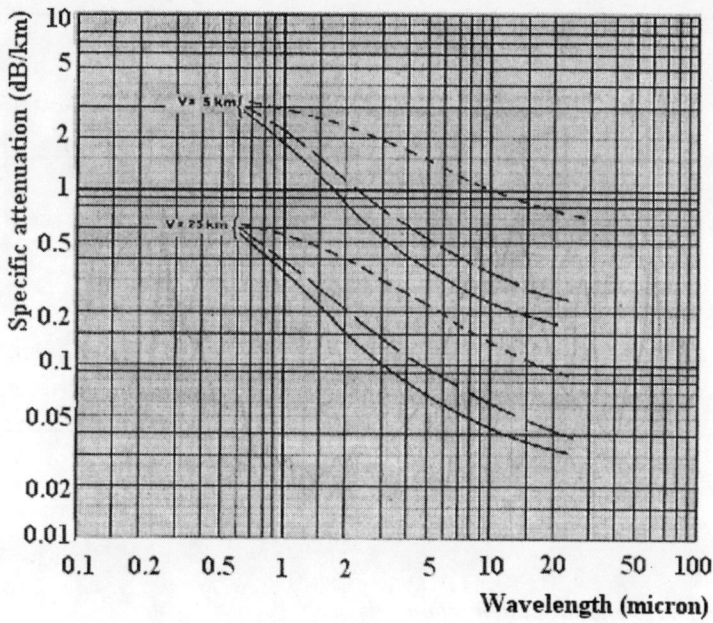

Figure 4.14: *Attenuation coefficient due to aerosols, according to visibility [CCIR, 1990]*
——— *Rural model*
— — *Urban model*
--------- *Maritime model*

Type of cloud	Wavelength (μm)				
	0.5	1	3	5	10
Good weather cumulus	0.09	0.095	0.1	0.1	0.05
Stratocumulus	0.18	0.18	0.2	0.21	0.12
Nimbostratus (cloud of rain)	0.5	0.6	0.6	0.65	0.5

Table 4-7: *Values of attenuation according to the wavelength for various types of clouds (cumulus, stratocumulus and nimbostratus)*

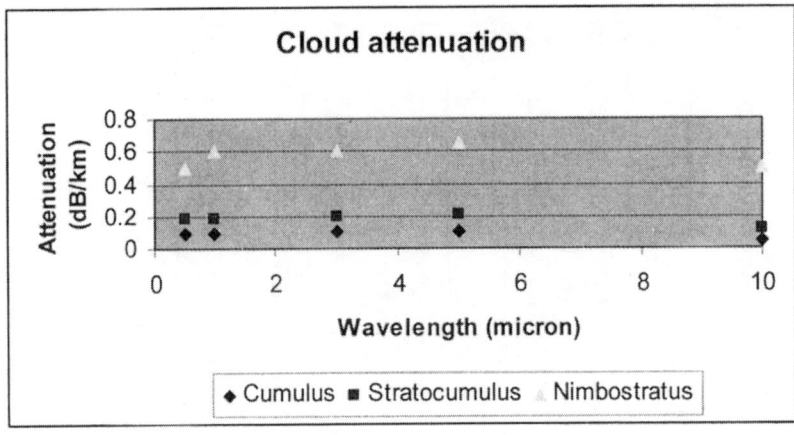

Figure 4.15: *Attenuation by clouds*

4.4.3. *Fog, haze and mist attenuation*

Fog, haze and mist consist of fine water droplets (<100 μm) suspended in the air. In certain very cold countries, they can be particles of ice. These droplets form when moist air is cooled below its dew point: the air becomes saturated and the water vapor contained in the air condenses in the form of fine water droplets.

To distinguish fog from haze and mist, it is generally agreed that the visibility is higher than 1000 m in haze or mist, whereas it is lower than 1000 m in fog. Their composition and size distributions vary strongly. In general, fogs have lower water contents than clouds (<0.1 g/m^3) and a smaller concentration of droplets (<100 cm^{-3}). It is characterized by the visibility given as the maximum distance beyond which a prominent object can not be seen by a human observer. The occurrence of fog is very variable, depending on geographical situation, proximity to the sea or a large lake, mountainous zones, etc. [LAVERGNAT, 1997].

Figure 4.16 below shows the general tendency of fog and mist attenuation according to wavelength and visibility distance. The curves are, in the main, based on calculation but agree reasonably well with experimental results [CHU, 1968].

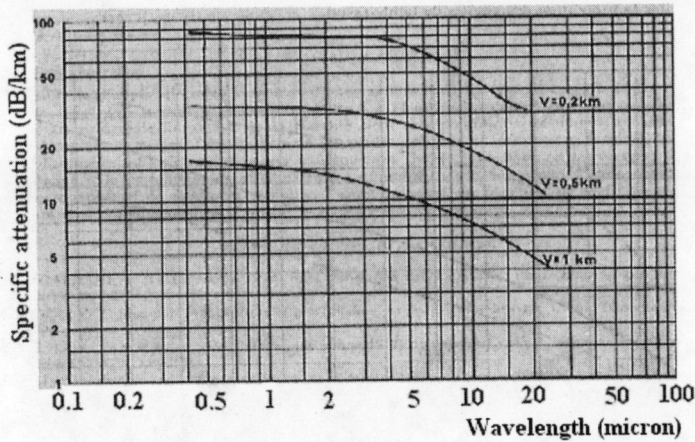

Figure 4.16: *Specific attenuation due to fog (V<1km) and to mist (V>1km), in relation to wavelength, for different values of visibility*

In addition, excellent correlation was found between attenuation due to fog and the liquid water concentration. The following specific attenuation coefficients were obtained at 0.63 µm and 10.6 µm [VASSEUR, 1997]:

$$\gamma_{0,63\eta m} = 360W^{0,64} \qquad\qquad [4.25]$$

$$\gamma_{10,6\eta m} = 610W \qquad\qquad [4.26]$$

where W is the liquid water concentration (g/m^3).

The liquid water concentration in fog is typically equal to approximately 0.05 g/m^3 for a moderate fog (visibility of about 300 m) and to 0.5 g/m^3 for a thick fog (visibility of about 50 m) [UIT-R P. 840-3].

Note – meteorological measurements carried out in Belgium in 16 stations over about twenty years provide indications on the maximum annual frequency (worst case scenario) and median (not exceeded in 50% of the cases). Table 4-8 gives the annual frequency of these fog events [BODEUX, 1977].

	Median frequency	**Max. frequency**
Light, moderate or thick fog ($V < 1000$ m)	0.055%	0.14 %
Moderate or thick fog ($V < 500$ m)	0.035%	0.11 %
Thick fog ($V < 200$ m)	0.020%	0.08 %

Table 4-8: *Annual frequency of fog events in Belgium*

Figure 4.17 below shows the distribution in France of the number of days per year with fog (days during which there was, even temporarily, a reduction in visibility to less than one kilometer).

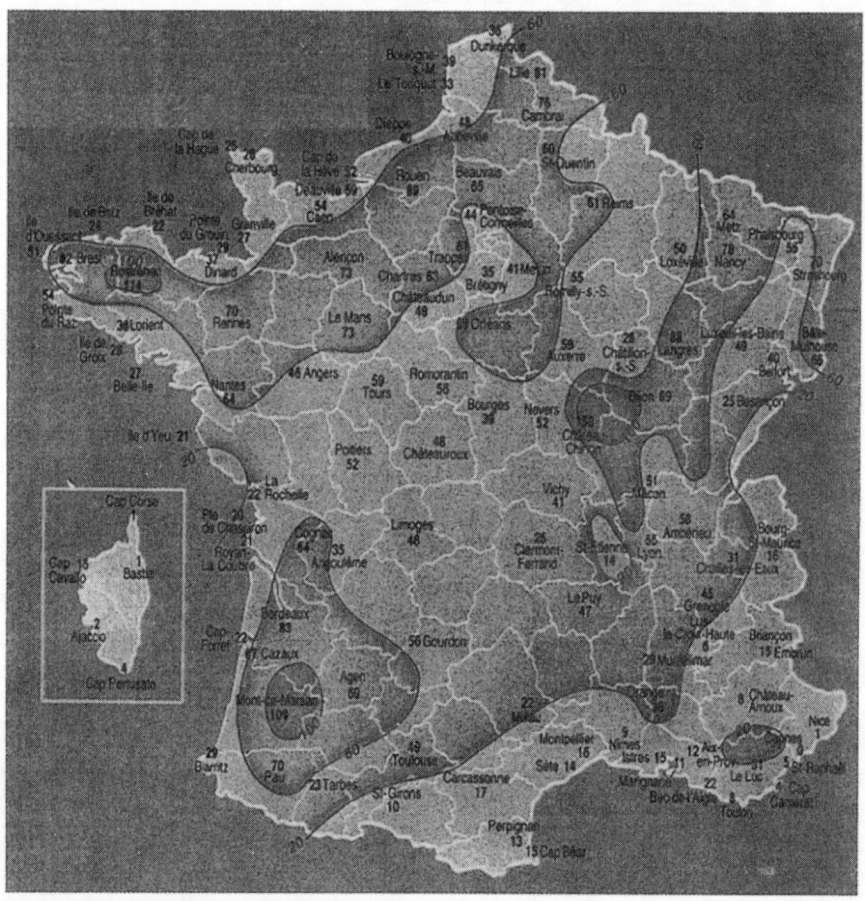

Figure 4.17: *Distribution in France of the number of days per year with fog (visibility lower than 1 km)*

4.4.4. *Precipitation attenuation*

4.4.4.1. *Rain*

Rain is formed from the water vapor contained in the atmosphere. It consists of water droplets whose form and number are variable in time and space. Their form depends on their size: they are considered as spheres until a radius of 1 mm and beyond that as oblate spheroids: flattened ellipsoids of revolution. Generally, the equivalent radius is used in calculations, i.e. the radius of the sphere that has the same volume.

Rain attenuation is due primarily to the scattering phenomenon as in the case of aerosols. In infrared, the wavelength is much smaller than the diameter of the raindrops. The value of the standardized cross section Q_d remains equal to 2 whatever the wavelength (geometrical optics field). Scattering attenuation depends of the quantity of raindrops intercepted in the radiation path.

The expression for the rain scattering coefficient is:

$$\alpha_n(\lambda) = 10^5 \int_0^\infty Q_a\left(\frac{2\pi r}{\lambda}, n''\right).\pi r^2 \frac{dN(r)}{dr} dr \qquad [4.27]$$

where r and $dN(r)$ are given in cm and in number/cm^4 respectively.

As the scattering cross section is equal to 2, the above expression becomes:

$$\alpha_n(\lambda) = 10^5 \int_0^\infty 2\pi r^2 \frac{dN(r)}{dr} dr \qquad [4.28]$$

The size distribution (characterized by the equivalent radius) results from complex processes like coalescence or bursting. The most commonly used distribution is that established by Marshall and Palmer (M-P distribution) [MARSHALL et al., 1943]:

$$N(r) = N_0 e^{-\gamma r} \qquad [4.29]$$

where:

– $N(r)dr$ is the number of water drops per unit volume whose equivalent radius lies between r and $r + dr$.

– N_0 (in m^{-3} mm^{-1}) and γ (in mm^{-1}) are experimentally determined constants; they depend on the type of rain under consideration.

The following values are usually accepted:

– $N_0 = 16*10^3$ m^{-3} mm^{-1};

– $\gamma = 8.2\ R^{0.21}$ mm^{-1} where R is the rain intensity in mm/h.

When the size of the irregularities due to precipitation becomes important compared to the wavelength, the wave is attenuated by reflection and refraction. Attenuation, independent of the wavelength, is a function of the rain intensity R (in mm/h) according to the following equation:

$$A_{precipitation}(dB/km) = aR^b \qquad [4.30]$$

where:

   $-a = 0.365$

   $-b = 0.63$

   $-R$ is the rain intensity in mm/h.

M-P distribution is only an average distribution and the use of high frequencies requires more elaborate analysis. M-P distribution underestimates very small drops on the one hand (drizzle) and storms on the other hand. Joss et al. [JOSS, 1968] propose values of N_0 and γ according to the rain type: convective rain or stormy shower, continuous rain, drizzle. The values of the parameters N_0 and γ are then:

   $-N_0 = 28*10^2 \, m^{-3} mm^{-1}$; $\gamma = 6 \, R^{0.21} \, mm^{-1}$ for convective rain,

   $-N_0 = 14*10^3 \, m^{-3} mm^{-1}$; $\gamma = 8.2 \, R^{0.21} \, mm^{-1}$ for continuous rain,

   $-N_0 = 6*10^4 \, m^{-3} mm^{-1}$; $\gamma = 11.4 \, R^{0.21} \, mm^{-1}$ for drizzle.

The corresponding values of the parameter a in the relation $A_{precipitation} = aR^b$ are described in Table 4-9 below [BATAILLE, 1992]:

Rain intensity (mm/h)	Type of rain	a
$R < 3.8$	Drizzle or light rain	0.509
$3.8 < R < 7.6$	Mean rain	0.319
$R > 7.6$	Convective rain	0.163

Table 4-9: *Influence due to the type of rain*

For high intensities of rain (particularly storm), the Law and Parson distribution, [LAW, 1943] gives specific attenuation coefficients more accurately than those given by Joss et al. Distribution models of a more complex decreasing exponential form have been advocated by different authors: [BEST, 1950, KHRGIAN, 1952, FUJIWARA, 1960]. Despite being among the oldest, the Law-Parson distribution and the Marshall-Palmer decreasing exponential and Joss et al. distribution still remain most commonly used for the calculation of rain attenuation [OLSEN, 1978]. These models were established based on experimental observations carried out in temperate areas.

The scattering properties of water drops depend not only on their shape but also on the characteristics of water. The complex permittivity, and therefore the refractive index, varies with frequency following Debye theory. From the communication systems point of view, rain is characterized by a statistical description of the disturbances induced by hydrometeors on a link determined in space and time. The rainfall rate R, measured in mm/h, is the fundamental parameter used to describe the rain locally. Its measurement is carried out either directly at ground level using pluviometers or comparable apparatus with optimum integration time of 1 minute, or indirectly using weather radar. The latter is particularly well adapted to rain structure analysis. Two types of rain are distinguished:

– wide or stratiform precipitation has a large spatial extension (a few hundred kilometers), its duration at a given location is long (several hours) and its intensity is moderate (< 25 mm/h). The rainfall rate is not uniform, even if the space variation is small. Radar measurements reveal a vertical homogeneity up to 0°C. Above this temperature, a layer, known as a mixture layer, is detected, in which ice, sleet or slush (melted snow) and complex aggregates coexist. This thin layer is distinctly visible in reflection; it is called the brilliant band. It has no influence on absorption at centimeter frequencies, but at millimeter frequencies it should probably be taken into account. The 0°C isotherm level is a fundamental parameter for oblique links and for this type of rain. An average height of the 0°C isotherm is given in Rec. UIT-R P. 839.

– convective precipitation, generally associated with meteorological fronts, has a small spatial extension, short duration (of the order of minutes) and very high rate of rainfall. Storms are a spectacular example of this. A considerable vertical extension, extending beyond the isotherm 0°C, is observed facing the front (hence the term convective) due to strong vertical air movements. In the back, a situation closer to stratiform rains is observed; the brilliant band is not always present. The horizontal variability of rainfall rate is very strong. There is the presence of very high rain intensity in isolated areas (>100 mm/h) called "rain cells", whose characteristic dimensions can range from 500 m to 10 km for "super-cells". The knowledge of the size of these cells is obviously essential in order to calculate attenuation.

In addition to space distribution, information about the occurrence (time distribution) is necessary to complete rain modeling. The exceeding probability, often expressed as a percentage, is the probability that a rainfall rate R is exceeded or equaled. For example, if the exceeding probability is 0.1%, that means that the rainfall rate R will be exceeded for 0.012 month per year, that is to say approximately 8 hours. UIT-R (Rec. UIT-R P.837) gives world charts of the rainfall rate exceeded for any given percentage of the average year with an integration time of 1 minute. However, these standard distributions do not pretend to replace, when they exist, precise measurements taken locally over long periods of time [LAVERGNAT, 1997].

The relation given by Carbonneau *et al.* [CARBONNEAU, 1998] is different; the attenuation values (dB/km) are much higher:

$$A_{precipitation} = 1.076 \times R^{0.67}$$ [4.31]

Figure 4.18 shows the variations of the specific attenuation (dB/m) due to precipitation in the optical and infrared wave range.

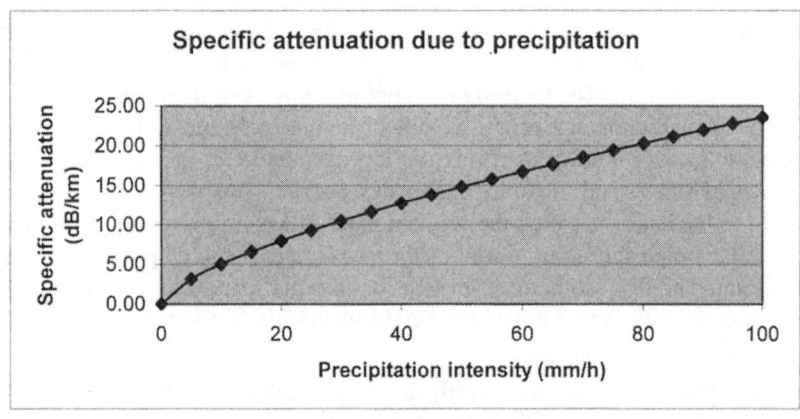

Figure 4.18: *Specific attenuation (dB/m) due to precipitation in the optical and infrared wave range*

Figure 4.19 shows the distribution of the precipitation intensities observed on the site of Fontaine (France) as well as a comparison with the ITU-R model. The model behavior is optimistic overall since it underestimates measurements on all the precipitation intensities range. This difference between predictions and measurements can be explained by the relatively low number of samples of treated data (only 20 months): variability from one year to another is not taken enough into account in such a short duration. The results over 10 years (desirable period for precipitation intensity measurements) for example, should certainly be different [VEYRUNES, 2000].

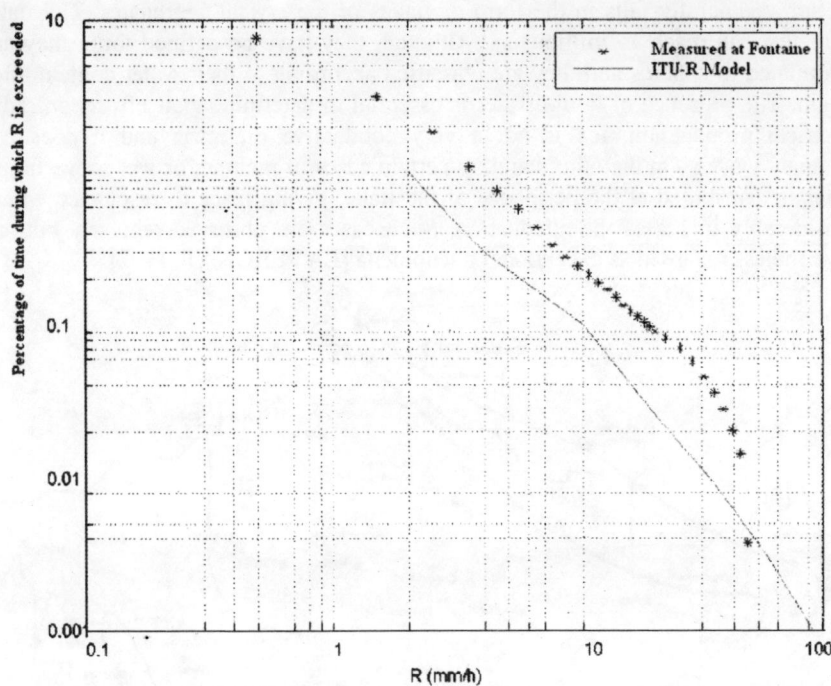

Figure 4.19: *Distribution of the intensities of precipitation observed on the site of Fontaine (France) as well as a comparison with ITU-R model*

Attenuation	Relation	980 and 1550 µm
Drizzle or light rain ($R < 3.8$)	$0.509\,R^{0.63}$	0.79 dB/km ($R = 2$)
Mean rain ($3.8 < R < 7.6$)	$0.319\,R^{0.63}$	0.88 dB/km ($R = 5$)
Strong rain (storm) ($R < 7.6$)	$0.163\,R^{0.63}$	1.08 dB/km ($R = 20$)
Rain (Marshall and Palmer)	$0.365\,R^{0.63}$	1.56 dB/km ($R = 10$)
Rain (Carbonneau)	$1.076\,R^{0.67}$	5.06 dB/km ($R = 10$)

Table 4-10: *Relation between the attenuation and the precipitation rate*

Figure 4.20 shows, for France, the curves of equal rain intensity in one minute exceeded during 0.01% of the time in an average year [ROORYCK, 1985].

4.4.4.2. *Other hydrometeors (snow, hail, etc.)*

Others hydrometeors are present in nature. They are mixtures of ice, air and/or liquid water. Among these mixtures, snow is one of most complex.

Snow generally falls in the form of flakes or ice crystal aggregates. The flake diameters can reach 15 millimeters. Although they have no defined form, they are assimilated to spheres and they are classified according to their water content after fusion. The presence of liquid water is essential to determine their effects on radio electrical propagation. Ice is not a very conductive dielectric and it does not attenuate waves. On the other hand, in certain cases of melting, or wet snow, liquid water is distributed at the periphery of the flake, giving place to resonance which considerably increases the attenuation. Such cases are observed regularly but are nevertheless regarded as exceptional phenomena [LAVERGNAT, 1997].

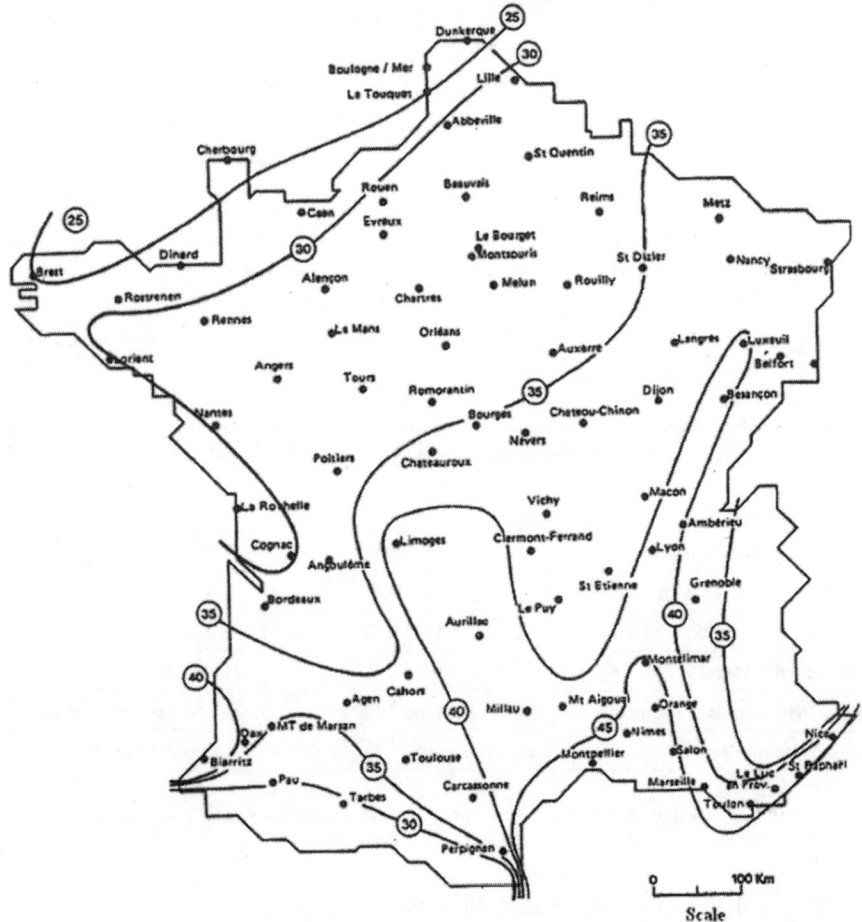

Figure 4.20: *Curves of equal rain intensity in one minute in France exceeded during 0.01% of the time on an average year [ROORYCK, 1985]*

Attenuation is strongly related to humidity (or the density of snow) and to the real snow rate S (mm/h). Experimental data analysis [VASSEUR, 1997] gives the following results:

− for low humidity snow falls (dry snow), attenuation can reach 20 dB/km in the visible region and 40 dB/km in the infrared region,

− for wet snow falls, attenuation varies from 4 to 8 dB/km for both visible and infrared waves.

The flake size distribution function (number of flakes whose size lies between r and $r + dr$ in m^{-3} mm^{-1}) is written:

for dry snow:

$$dN(r)/dr = 10^9 r^2 \exp(-46.30 \, S^{-0.28} \, r) \tag{4.32}$$

for wet snow:

$$dN(r)/dr = 12*10^6 r^2 \exp(-20.0 \, S^{-0.15} \, r) \tag{4.33}$$

The real snow fall rate S (mm/h) can be deduced using the expression proposed in the literature [O'BRIEN, 1970], which is based on size distribution functions and flake final speed. The associated precipitation rate is connected to the snow fall rate through the ratio of the density of snow compared to water. The forward-scattering and the multipath effects are not very important in infrared but are very marked in the visible spectrum.

The relation between the attenuation before saturation and snow fall rates can be described by a law of the type:

$$Aff_{snow} \, [dB/km] = aS^b \tag{4.34}$$

The a and b coefficients are mentioned in Table 4-11. Figures 4.21, 4.22 and 4.23 show the attenuation variations for various conditions of snow fall, for the 0.63 and 10.6 µm wavelengths.

		10.6 µm	0.63 µm
Wet snow	a	4.87	3.85
	b	0.75	0.72
Dry snow	a	6.07	5.53
	b	1.40	1.38

Table 4-11: *Numerical values of the a and b coefficients allowing evaluating snow attenuation for the 0.63 and 10.6 µm wavelengths*

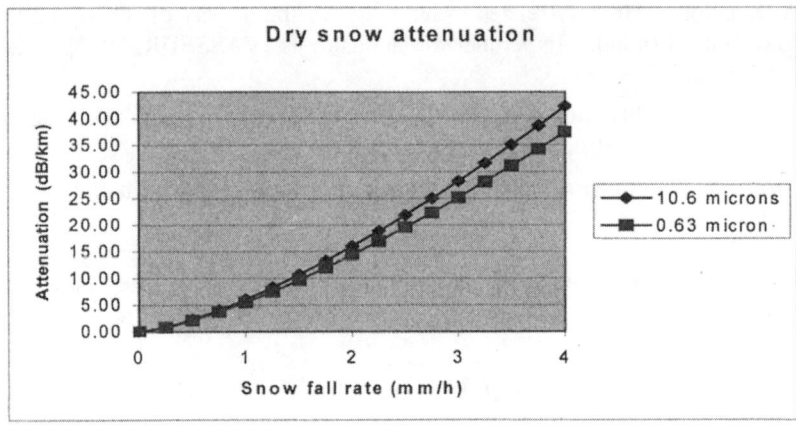

Figure 4.21: *Attenuation due to dry snow according to snow fall rate (mm/h) for the 0.63 and 10.6 μm wavelengths*

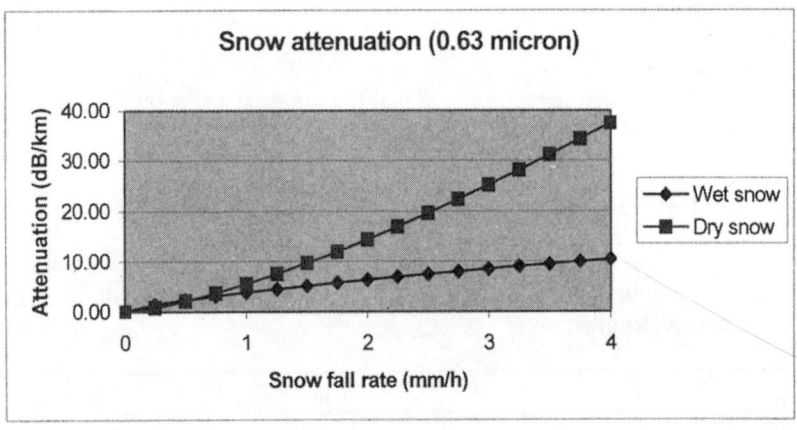

Figure 4.22: *Comparison of the attenuation due to wet and dry snow according to snow fall rate (mm/h) for the 0.63 μm wavelength*

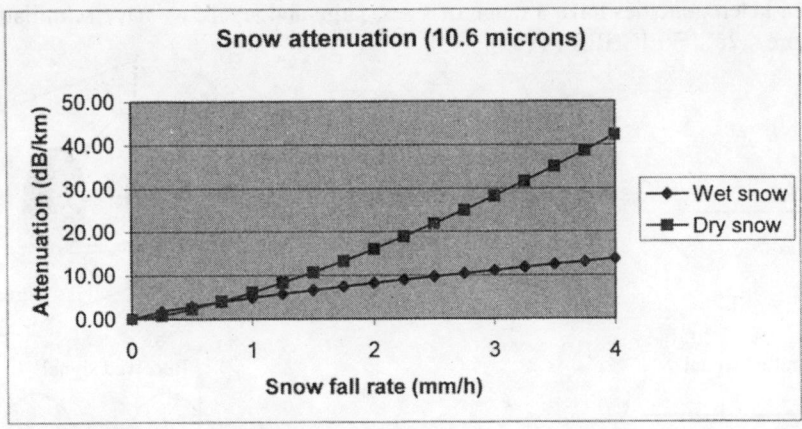

Figure 4.23: *Comparison of the attenuation due to wet and dry snow according to snow fall rate (mm/h) for the 10.6μm wavelength*

Hail can be regarded as ice containing air bubbles. Hailstones are the largest observable hydrometeors (up to 8 cm in diameter). Made of ice, hailstones do not attenuate waves a lot, at least in the centimeter range. On the other hand, for millimeter waves and at higher frequencies the scattering cross section becomes important because it can lead to an awkward extinction.

This phenomenon has a very localized nature and is therefore neither very common nor very alarming. Its occurrence is not well known [LAVERGNAT, 1997].

4.4.5. *Refraction and scintillations*

Randomly distributed cells are formed under the influence of thermal turbulence inside the propagation medium. They have variable size (10 cm–1 km) and differ in temperature. These cells have different refractive indices thus causing scattering, multipaths, and variation of the incident angles: the received signal fluctuates quickly at frequencies ranging between 0.01 and 200 Hz. The wave front varies similarly causing focusing and defocusing of the beam. Such fluctuations of the signal are called scintillations. The amplitude and the frequency of scintillations depend on the size of the cells compared to the beam diameter. The following figures schematize this effect as well as the variations (amplitude, frequency) on the received signal. When hetcrogcneities are large compared to the beam cross section, it is deviated (Figure 4.24), when they are small, the beam is widened (Figure 4.25).

When heterogeneities have a range of sizes, large and small, we have scintillations (Figure 4.26) [WEICHEIL, 1989].

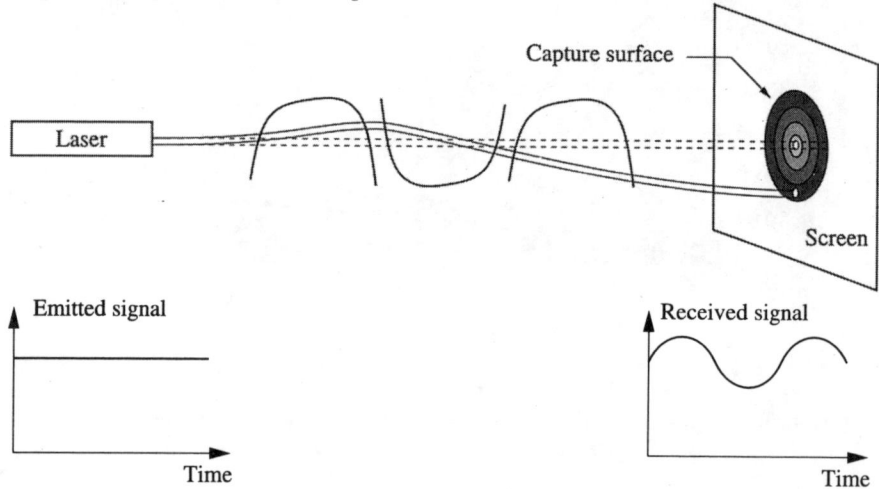

Figure 4.24: *Deviation of a beam under the influence of turbulence cells larger than the beam diameter*

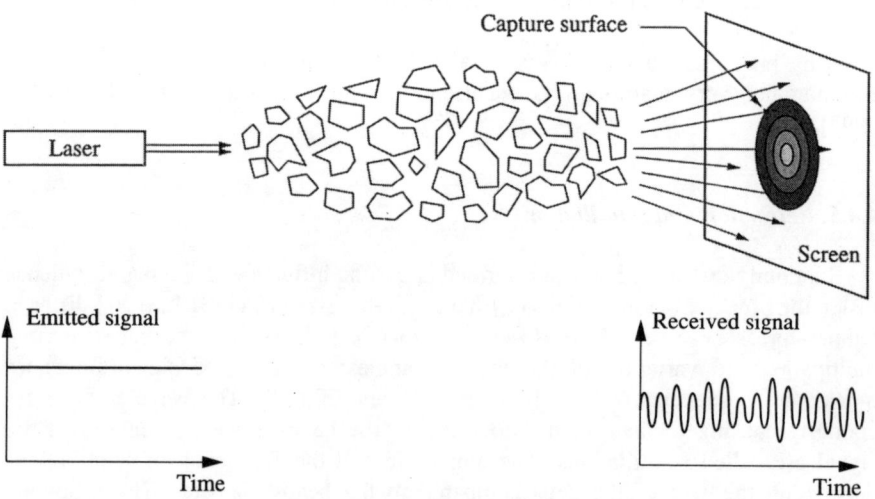

Figure 4.25: *Deviation of a beam under the influence of turbulence cells smaller than the beam diameter of the beam (widening of the beam)*

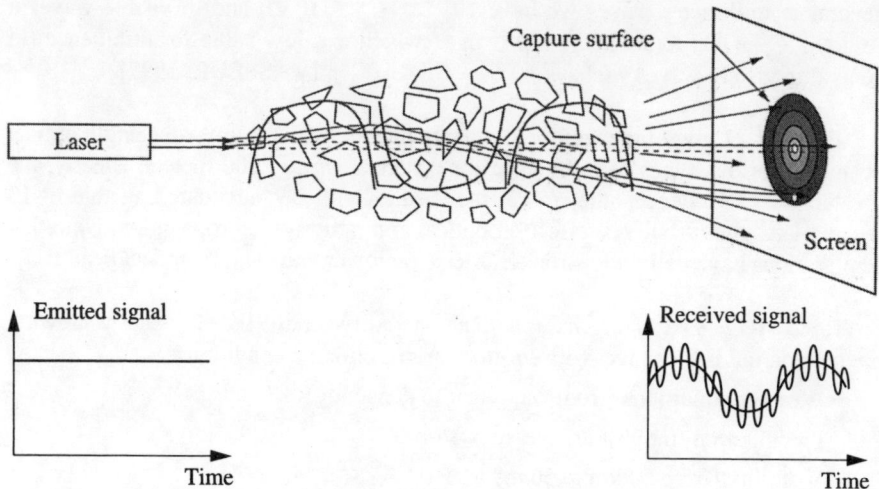

Figure 4.26: *Effects of various different sized heterogeneities on a laser beam propagation (scintillations)*

Tropospheric scintillation effects are generally derived from the logarithm of the amplitude χ [dB] of the observed signal ("log-amplitude"), defined as the ratio, in decibels, between its instantaneous amplitude and its average value. The intensity and the speed of the fluctuations (scintillation frequency) increase with wave frequency. For a plane wave, a low turbulence and a specific receiver, the scintillation variance σχ2 [dB²] can be expressed by the following relation:

$$\sigma_\chi^2 = 23.17 * k^{7/6} * C_n^2 * L^{11/6} \qquad [4.35]$$

where:

– k [m^{-1}] is the wave number ($2\pi/\lambda$),

– L [m] is the length of the link,

– C_n^2 [m$^{-2/3}$] is the structural parameter of the refractive index, representing the intensity of the turbulence.

Scintillation peak-to-peak amplitude is equal to $4\sigma_\chi$ and attenuation related to scintillation is equal to $2\sigma_\chi$. For strong turbulence, a saturation of the variance given by the above relation is observed [BATAILLE, 1992]. The parameter C_n^2 does not have the same value for both millimeter waves and optical waves [VASSEUR, 1997]. Millimeter waves are especially sensitive to humidity fluctuations while in the visible range, refractive index is primarily a function of the temperature (the water vapor contribution being negligible). One obtains in millimeter waves a value of C_n^2 equal to about 10^{-13} m$^{-2/3}$ which is an average value for turbulence (in general in millimeter waves we have $10^{-14} < C_n^2 < 10^{-12}$) and in visible waves a value of C_n^2 equal to about 2×10^{-15} m$^{-2/3}$ which is a low value for turbulence (in general in visible waves we have $10^{-16} < C_n^2 < 10^{-13}$), [VASSEUR, 1997].

Figure 4.27 gives the variation of the attenuation of a 1.5 μm wavelength optical beam for various types of turbulence for distances up to 2000 meters. Clearly, the greater the turbulence, the more the beam will be attenuated. Table 4-13 recapitulates the turbulence effect on optical and radio wave propagation. Note that scintillations have definitely stronger effects on lower wavelength optical beams.

Table 4-12 gives the international visibility code [KIM, 1998] showing attenuation in visible waves (dB/km) for various climatic conditions:
- Weather conditions (from very clear to dense fog)
- Precipitation (mm/h): drizzle, rain, storm
- Visibility (from 50 km to 50 m)

International visibility code				
Weather conditions	Precipitation	mm/h	Visibility (m)	Attenuation (dB/km)
			0	
Dense fog			50	315
Thick fog			200	75
Moderate fog			500	28.9
Light fog	Storm	100	770	18.3
			1,000	13.8
Very light fog	Strong rain	25	1,900	6.9
Snow			2,000	6.6
Light mist	Average rain	12.5	2,800	4.6
			4,000	3.1
Very light mist	Light rain	2.5	5,900	2
			10,000	1.1
Clear air	Drizzle	0.25	18,100	0.6
			20,000	0.54
Very clear air			23,000	0.47
			50,000	0.19

Table 4-12: *International visibility code [KIM, 1998] showing attenuation in the visible waveband (dB/km) for various climatic conditions*

	Turbulence		
	Low	moderated	High
C_n^2 optic waves	10^{-16}	10^{-14}	10^{-13}
Attenuation (0.98 µm)	0.51	5.06	16.00
Attenuation (1.55 µm)	0.39	3.87	12.25
C_n^2 millimeter waves	10^{-15}	10^{-13}	10^{-12}
Attenuation (40 GHz)	0.03	0.09	0.27
Attenuation (60 GHz)	0.03	0.11	0.35

Table 4-13: *Summary table of attenuation due to scintillations*

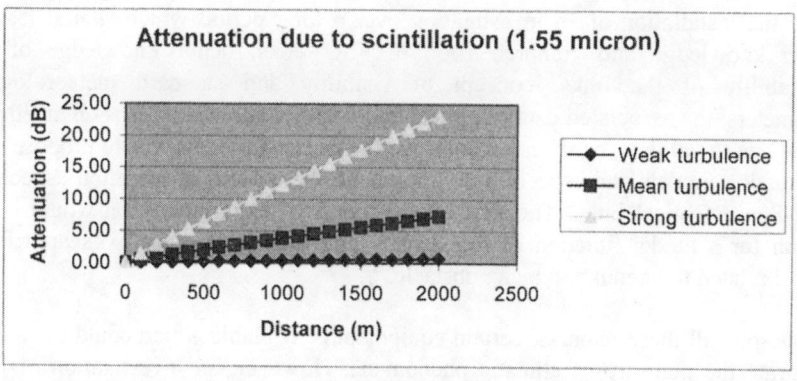

Figure 4.27: *Variation of attenuation due to scintillation according to distance for various types of turbulence at 1.55 µm*

4.5. Free-Space Optical links

In this chapter, some of the aspects of propagation to be considered at the point of installation of Free-Space Optical (FSO) links were presented (for example, atmospheric effects such as molecular and aerosol absorption, molecular and aerosol scattering, extinction, cloud attenuation, rain attenuation, refraction and scintillations). Weather phenomena bring the most important attenuation and particularly: fog, dry snow, rainstorms (generally short) and light rain (in this order of importance). The fog constitutes the most penalizing phenomenon. It occurs frequently, certain years and is slow to dissipate. Because of the intensity of dry snow, FSO equipment cannot currently be used in mountainous zones.

Currently, there is little in the form of documentation, recommendations, standards or software allowing a budget calculation of links using data related to availability and quality of service. Three approaches are to be recommended in order to obtain a model for link budget calculation following the example of radio-relay systems (UIT-R P.530-8) *"Propagation data and prediction methods required for the design of terrestrial line-of-sight systems"*:

− a more detailed analysis of the collected documents, taking as the first step Recommendation UIT-R P.530-8, in order to outline a model which would not be, initially, a very good approximation, because the validity of the suggested model will be limited by the lack of historical data,

− a complete information search for the various fog densities (liquid water concentration in g/m^3) and their annual frequencies (expressed as a percentage) over a given geographical area in order to determine the percentage of time for which the link margin is exceeded,

− the installation of an investigation over a long period which should lead to better knowledge and comprehension of FSO links, better knowledge of the availability of the links, concept of visibility and standard meteorological parameters; the associated data of optical and weather attenuation thus obtained will allow one to validate, confront, or improve the existing models, and to propose new attenuation models and especially to apprehend the various propagation aspects of Free-Space Optical links. The first two elements of experimentation would be: i) search for a model function of fog density and time percentage, ii) search for a model related to attenuation by an obstacle.

Despite all these remarks, certain equipment is available which could be used to integrate the majority of climatic phenomena. However, an investigation of FSO equipment is a prerequisite, in order to validate conformity to expectations of a given quality of service. If a base of availability borrowed from radio-relay system is taken, FSO equipment should reach 99.999%. In reality, these percentages of time are slightly smaller. A software tool to simulate the quality of service of a free space optical link was developed by FTR&D. It allows, on a given geographical site, the determination of the availability and reliability of a link according to systems (power, wavelength, characteristics of the material), climatic and atmospheric parameters. It integrates various physical phenomena responsible for link rupture such as attenuation due to ambient light, scintillation, rain, snow and fog [CHABANE, 2004].

Finally, it appears that attenuation phenomena caused by the atmosphere, clouds, fog and scintillation, are least significant at wavelengths of about 10 or 20 μm. However, at these wavelengths, Free-Space Optical equipment does not yet exist.

Chapter 5

Propagation of an Optical Beam
in Confined Space

5.1. Introduction

The conjugation of an unceasingly increasing request rate for communications data and the high cost of installation and maintenance of wire networks inside buildings, leads us to turn towards more flexible and economic solutions, namely wireless technologies. Distinguished among these are radio electrical and infrared techniques.

The latter is particularly interesting for various reasons: frequency broadband availability at low cost, worldwide unregulated frequency spectrum, immunity against radio electrical interference, availability of components coming from optics, low consumption, etc.

In order to better evaluate the performance and to optimize the deployment of such communication systems at a high data rate in confined space (interiors of buildings, halls, conference rooms, libraries, hospitals, newspaper offices, etc.) good knowledge of the various mechanisms of propagation is necessary. This chapter is essentially devoted to this. It considers the principal propagation mechanisms (line of sight (LOS), wide-LOS or cellular, reflected or diffuse links); the study of the propagation channel, its modeling; the power necessary to reach a requested binary rate; the various sources of optical noise and its modeling.

5.2. Various mechanisms of propagation

The propagation paths between the transmitting source and the receiving cell (Figure 5.1) are different in nature. The first distinguishes the Line Of Sight (LOS) link (Figure 5.1a), the second the wide-LOS or cellular link connection (Figure 5.1b), the third the specular reflection link (Figure 5.1c) and fourth the diffuse reflection link (Figure 5.1d). Two hybrid systems combine different combinations of these mechanisms (Figures 5.1e and f).

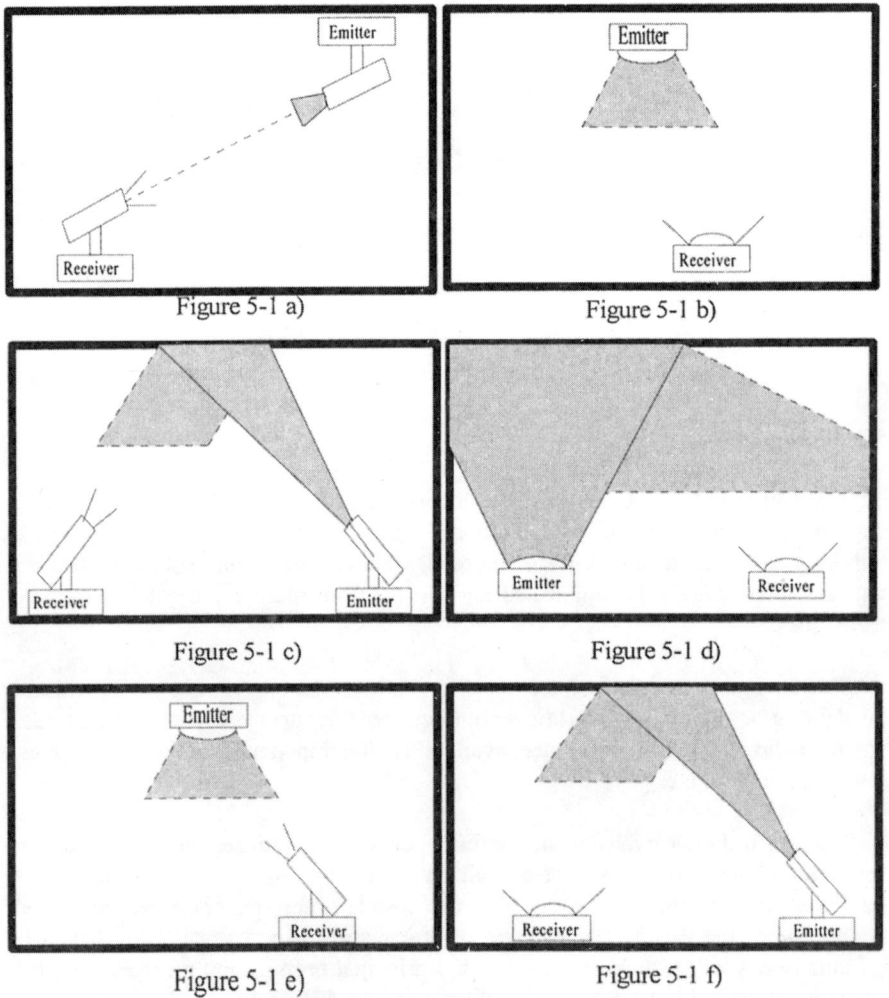

Figure 5-1 a)

Figure 5-1 b)

Figure 5-1 c)

Figure 5-1 d)

Figure 5-1 e)

Figure 5-1 f)

Figure 5.1: *Various configurations of infrared links*

5.2.1. *Line Of Sight links*

Line Of Sight (NLOS) systems, generally of low power, concentrate their energy in a very narrow beam, thus allowing a high power flux density at the receiver. The beam emitted by the transmitter is, however, only slightly divergent, so the receiving cell will only collect a fraction of the emitted power. From simple geometrical consideration, the free-space loss undergone by such systems can be represented by the following equation, for small angles [TRAVIS, 1996, STREET, 1997]:

$$Aff \approx 20\log_{10}\left(\frac{2\theta d}{D_{RX}}\right)$$

[5.1]

where:

– d is the distance between the transmitter and the receiver,

– D_{RX} is the diameter of the receiver capture surface,

– θ is the half-angle beamwidth.

Systems operating according to this principle present certain advantages but also certain disadvantages.

Advantages of such systems are:

– only the direct path is considered; the link is not disturbed by the presence of multiple paths, the receiver does not need to have a large field of view (FOV),

– the effective gain of the concentrator can be exploited to improve the link budget,

– narrow band thin film optical filters can be used because the angular dependence of the filter response is not an issue [BARRY, 1995].

Disadvantages of such systems are:

– an accurate alignment between the transmitter and the receiver is required,

– they can be subject to cuts and blockages, particularly due to the displacement of the person in the room or to the presence of an obstacle between the transmitter and the receiver.

The most successful standard adopted to date is the Infrared Data Association (IrDA). IrDA is an "industrial club" formed in 1993 by Hewlett-Packard and IBM. Currently it consists of more than 100 computing organizations such as Apple, Microsoft, and Intel. The aim of IrDA is to define a half-duplex serial data interconnection standard, at low cost, low power and ensuring interoperability between the various systems.

5.2.2. *Wide-LOS or cellular links*

Wide-LOS systems allow the potential coverage area to be increased. Hence, the power flux density at the receiver is lower in comparison to a LOS system working under identical conditions (same emitted power, same distance).

A typical configuration consists of the base station located in the ceiling in order to provide coverage of a mobile station within a cell. Such systems allow telepoint connections [NICHOLLS, 1996] and cellular coverage of a large area [SMYTH, 1993, SMITH, 1995, McCULLAGH, 1994].

The worst-case losses at the edge of the coverage, assuming a uniform illumination inside a cone with half-angle θ_{max}, the receiver being aligned parallel to the ceiling (see Figure 5.2), are given by the following approximate relation [STREET, 1997]:

$$Aff_{(max)} \approx \frac{A_{RX} \cos^3 \theta_{max}}{2\pi (h-x)^2 (1-\cos\theta_{max})} \qquad [5.2]$$

where:

– h is the height of the room,

– x is the height of the coverage area above the floor,

– A_{RX} is the capture area of the receiver.

For example, for a room with the following dimensions: length = 4 m, width = 4 m and height = 4 m and $X = 1$ m, then $\theta_{max} \approx 54°$ and the path loss at the edge of the coverage is in the order of 57 dB.

The interference potential due to the presence of multiple paths increases with the beamwidth of the transmitter.

Such a configuration can be extended, with a cellular system, to the coverage of a room by using a number of base stations (access points) located in the ceiling.

However, such systems suffer blocking problems. These can be mitigated by using several base stations located in different points. In this case, system complexity and cost will increase accordingly.

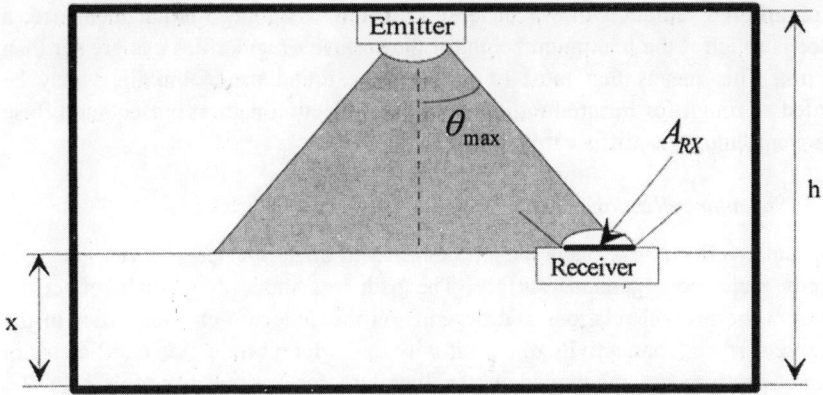

Figure 5.2: *Geometrical configuration of a wide-LOS link*

5.2.3. *Reflected links*

Reflection occurs when the wave meets a surface whose dimensions are large compared to the wavelength (floor, wall, ceiling, furniture, etc.).

The reflection characteristics of a surface depend on several factors, in particular the surface material (smooth or rough), the wavelength of the incident radiation and the angle of incidence.

The roughness of surface relative to the wavelength constitutes an important parameter of the shape of the reflection pattern. A smooth surface reflects the incident radiation in a single direction like a mirror (specular reflection). A rough surface, on the other hand, will reflect the incident radiation in several directions.

A surface is considered as rough, according to the Rayleigh criterion, if the following relation is satisfied:

$$\varsigma > \frac{\lambda}{8\sin\theta_i} \qquad [5.3]$$

where:

- ς is the maximum height of the surface irregularities,
- λ is the wavelength of the incident radiation,
- θ_i is the angle of incidence.

For infrared radiation of wavelength 1550 nm, assuming normal incidence, a surface is rough if the maximum height of the surface irregularities ç is greater than 0.19 μm. This means that most of the surfaces found inside buildings may be regarded as rough for infrared radiation, so the reflection pattern presents a diffuse component (known as diffuse reflection).

5.2.3.1. *Specular reflection*

Specular reflection, a phenomenon common to all frequencies, is that due to a perfectly plane homogeneous surface. The path loss induced by such reflections rises from the Fresnel relations and depends on the dielectric characteristics of the reflective surface (conductivity σ, permittivity ε). Different reflection coefficients in infrared spectrum are mentioned in the literature for various materials [YANG, 2000] (Table 5-1).

Material	Reflection coefficient	Material	Reflection coefficient
Painted wall	0.184	Glass	0.0625
Floor tile	0.128	Brown bookshelf	0.0884
Red brick	0.047	Monitor screen	0.0704
Computer box	0.1018	White ceramic	0.0517

Table 5-1: *Reflection coefficients of materials or furniture usually found inside a room*

5.2.3.2. *Diffuse reflection*

Diffuse reflection is due to the reflections by irregularities on rough surfaces. It means that an incident wave is not simply reflected in a single direction but is diffused in multiple directions. Two models are usually used to represent the reflection of infrared radiation: the Lambert and the Phong models.

5.2.3.2.1. The Lambert model

Some surfaces are completely irregular and reflect the infrared radiation in all directions independently of the incident radiation. Such surfaces are known as diffuse and can be approximated using Lambert's model. This model is very simple and easy to implement using computational software. It is described by the following equation:

$$R(\theta_0) = \rho R_i \frac{1}{\pi} \cos(\theta_0) \qquad [5.4]$$

where:

    $- \rho$ is the surface reflection coefficient,

    $- R_i$ represents the incident optical power,

    $- \theta_0$ is the observation angle.

5.2.3.2.2. The Phong model

The reflection pattern of several rough surfaces is well approximated by Lambert's model except when close to the specular reflection where the reflection pattern presents an important component. The Phong model considers the reflection pattern as the sum of two components: the diffuse component and the specular component.

The percentage of each component depends mainly on the characteristics of the surface and is a parameter of the model. The diffuse component is modeled by the Lambert model. The specular component is modeled by a function which depends on the angle of incidence θ_i and the observation angle θ_0.

Phong's model is described by the following equation:

$$R(\theta_i, \theta_0) = \rho \frac{R_i}{\pi} \left[r_d \cos(\theta_0) + (1 - r_d) \cos^m (\theta_0 - \theta_i) \right] \qquad [5.5]$$

where:

    $- \rho$ is the surface reflection coefficient,

    $- R_i$ represents the incident optical power,

    $- r_d$ is the percentage of incident signal that is reflected diffusely (it is a value ranging between 0 and 1),

    $- m$ is a parameter which controls the directivity of the specular component of the reflection.

    $- \theta_i$ is the angle of incidence,

    $- \theta_0$ is the angle of observation.

The Phong model, which depends on the angle of observation and on the angle of incidence, is more complex than the Lambert model. The computing time required to simulate the infrared propagation for indoor environments is thus significantly higher. The reflection pattern presents a principal component centered on the direction of the specular reflection.

Many measurements of reflection patterns are found in the literature. Most emanate from Lomba et al. [LOMBA, 1998]; the authors present the reflection patterns of various materials usually used in indoor spaces in the form of polar diagrams: cement (before and after painting), wood (before and after varnishing), ceramic floor tiles, formica, white paper, brown pasteboard, glass, etc.): see Figure 5.3 below.

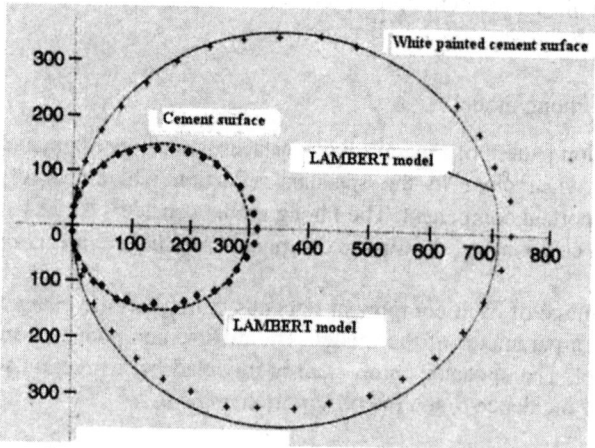

Figure 5.3a: *Experimental reflection patterns for an angle of incidence of 45° of a rough cement surface (before and after white painting) and their approximations according to the Lambert model*

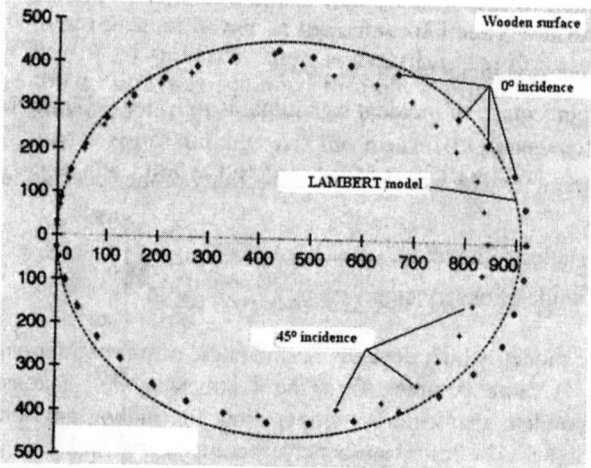

Figure 5.3b: *Experimental reflection patterns for angles of incidence 0°and 45°on a wooden surface and the approximation according to the Lambert model*

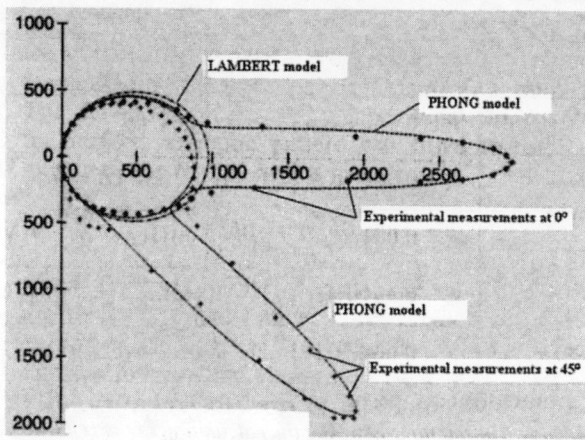

Figure 5.3c: *Experimental reflection patterns for angles of incidence of 0° and 45° on a varnished surface of wood, and their approximations according to the Lambert and Phong models*

Figure 5.3a shows the experimental reflection pattern on a rough cement surface (the most commonly used material in indoor space) on the one hand, and on the same surface covered with white paint on the other hand, for an angle of incidence of 45°. The two patterns are different; the white wall reflects a larger quantity of radiation but the shape of the pattern remains almost unchanged. The experimental values were approximated by the Lambert model. As the reflective surface is rough, the Lambert model correctly represents the experimental values.

Figure 5.3b shows the experimental reflection patterns on a plane surface of wood for angles of incidence of 0 and 45° as well as the associated Lambert model. This model is in excellent agreement with experimental measurements.

Figure 5.3c shows the experimental reflection patterns on a plane surface of varnished wood for angles of incidence of 0 and 45° as well as the associated Lambert and Phong models. The varnishing of the surface shows an important modification of the radiation pattern; an important specular component appears for each angle of incidence. This component is better taken into account by the Phong model than the Lambert model.

Reflection pattern models also exist in the literature. They result from the smoothing of experimental measurements carried out by many authors [NICODEMUS, 1977, PHONG, 1975]. We mention below the reflection coefficients relative to the following materials: painted wall, floor tiles, computer box, glass, brown bookshelf, monitor screen, white ceramic [YANG, 2000] (see Table 5-2).

Material	Reflection coefficient
Painted wall	$R = \cos(\theta_0)$
Floor tiles	$R = 0.6\cos(\theta_0) + (1-0.6)\cos^6(\theta_0 - \theta_i)$
Computer box	$R = 0.55\cos(\theta_0) + (1-0.55)\cos^3(\theta_0 - \theta_i)$
Glass	$R = 0.001\cos(\theta_0) + (1-0.001)\cos^{13}(\theta_0 - \theta_i)$
Monitor screen	$R = 0.39\cos(\theta_0) + (1-0.39)\cos^{10}(\theta_0 - \theta_i)$
White ceramic	$R = 0.06\cos(\theta_0) + (1-0.06)\cos(\theta_0 - \theta_i)$

Table 5-2: *Mathematical relations allowing the calculation of the reflection coefficients for various materials according to the angle of incidence θ_i and the angle of observation θ_0*

For example, the figures below show the variations of the relative reflection intensity of various materials with light incident normally (Figure 5.4) and at an angle of incidence of 30° (Figure 5.5) according to the angle of observation.

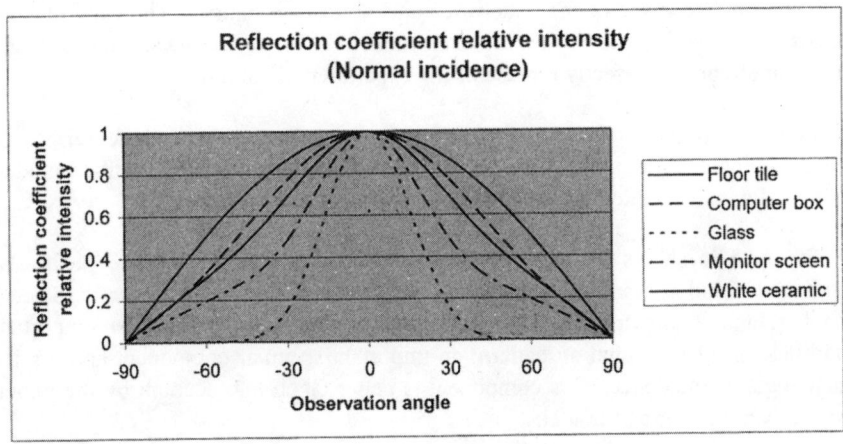

Figure 5.4: *Variations of the relative reflection intensity of various materials illuminated normally, according to the angle of observation*

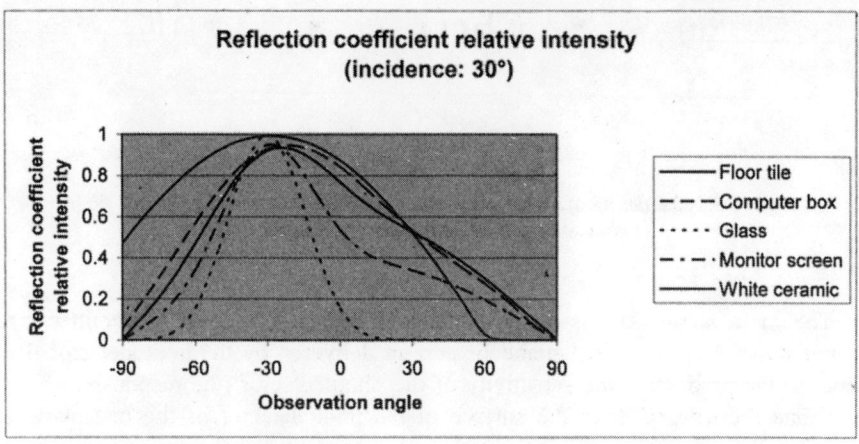

Figure 5.5: *Variations of the relative reflection intensity of various materials illuminated at an angle of incidence of 30° according to the angle of observation*

5.3. Propagation channel

5.3.1. *Description of an infrared propagation channel*

Figure 5.6 below describes an infrared link with intensity modulation and with direct detection (IM/DD: Intensity Modulation and Direct Detection). It consists of an infrared emitter (the transmitter) and a large area photodetector (the receiver).

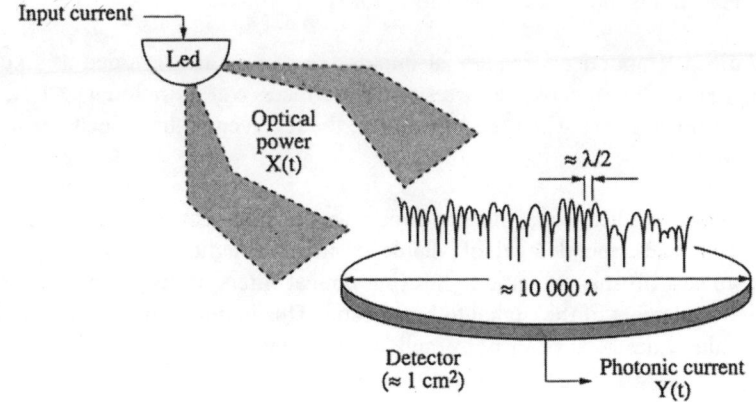

Figure 5.6: *Description of an infrared link*

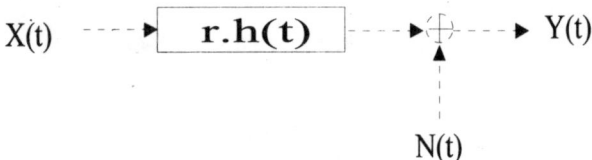

$$X(t) \dashrightarrow \boxed{\textbf{r.h(t)}} \dashrightarrow \oplus \dashrightarrow Y(t)$$

N(t)

Figure 5.7: *Modeling of an infrared channel by an time-invariant linear filter characterized by an impulse response h(t)*

The input signal $X(t)$ is the instantaneous optical power of the emitter. The output signal $Y(t)$ is the instantaneous current delivered by the photodetector. It is equal to the product of the sensitivity of the photodetector (the responsivity) R in A/W and the integral over the surface of the photodetector of the instantaneous optical power of each point. The signal propagates between the transmitter and the receiver through a room via possible reflections.

The channel is represented in baseband by a linear system characterized by the input power $X(t)$, the output intensity $Y(t)$, the additive white noise $N(t)$ and the impulse response $h(t)$ (Figure 5.7).

The propagation channel can be also described in term of frequential response by the following relation:

$$H(f) = \int_{-\infty}^{+\infty} h(t) e^{-2\pi i f t} dt \qquad [5.6]$$

where $H(f)$ is the Fourier transform of $h(t)$.

Various temporal or frequential impulse responses are regarded as fixed for a given transmitter, receiver and reflective surfaces configuration [KAHN, 1995]. These responses vary when the transmitter, the receiver or the objects are moved in the room.

In most applications, infrared links are carried out in the presence of a significant background level of visible or infrared light. Although it is possible partly to cast off this parasitic light using optical filters, it does, however, remain a restrictive element of the Signal to Noise ratio. This impulsive noise can be modeled as a white, quasi Gaussian noise and independent of the input signal $X(t)$ [LEE, 1994].

The propagation channel can thus be written in the following form:

$$Y(t) = rX(t) \otimes h(t) + n(t) \qquad [5.7]$$

where $\otimes$ indicates the product of convolution.

That relation can also be written as:

$$Y(t) = r \int_0^\infty h(\tau) X(t-\tau) d\tau + n(t) \qquad [5.8]$$

where $h(\tau)$ is the response at time t to a luminous impulse $X(t)$ which would have been emitted at time $t - \tau$. It allows the different echoes to be distinguished according to their propagation delay. It completely characterizes the propagation channel.

To represent the "indoor" channel in baseband, $h(t)$ is generally modeled as following:

$$h(t) = \sum_{n=1}^{n=N} a_n \delta(t-t_n) e^{j\theta_n t} \qquad [5.9]$$

where a_n, t_n, θ_n and N are statistically determined [HASHEMI, 1994].

The use of the Dirac function δ as a basic function is adapted in radio operator environments where the reflective elements are typically specular. In infrared, reflective surfaces are primarily diffuse, e.g. the reflected radiation is diffused in all directions according to a continuous and nearly independent distribution of angle of incidence [GFELLER, 1979]. Thus, when a reflecting object is illuminated by an emitting source and this object is in the field of view of the receiver, the emitted impulse extends in time. This leads us to consider, for the basic function, $h(t)$ impulse forms with duration and form corresponding to diffuse reflector responses.

5.3.2. Channel characterization

An infrared multipath channel is characterized to determine the transmitted average optical power, to allow some binary error rate (TEB) and for a given type of modulation.

As the input signal $X(t)$ represents the instantaneous optical power, the input channel is non negative ($X(t) \geq 0$). The transmitted average optical power is determined by the relation:

$$P_{tx} = \lim_{T \to \infty} \frac{1}{2T} \int_{-T}^{+T} X(t) dt$$

rather than by the temporal average of $|X(t)|^2$ which is suitable when $X(t)$ represents an amplitude.

The power necessary to reach the desired performances has two components:

– optical power lost in free space (optical path loss),

– power necessary to mitigate multipaths (multipath power requirement) [KAHN, 1997, CARRUTHERS, 1997].

Defining the optical gain G_0 for an infrared channel characterized by an impulse response $h(t)$ using the expression:

$$G_0 = \int_{-\infty}^{+\infty} h(t)dt$$

then the received optical power is defined by the relation $P_{rx} = G_0 P_{tx}$. So the power attenuation in free space is defined by the term: $-10\log_{10} G_0$.

The "multipath" aspect is deduced from parameters related to the impulse response of the propagation channel. The most important parameters are the mean delay and the root mean square (rms) delay spread.

The mean delay is the average of the delays weighted by their power. It is given by the following relation (first order moment of the impulse response):

$$\mu = \frac{\int t h^2(t)dt}{\int h^2(t)dt} \tag{5.10}$$

The rms delay spread or the standard deviation of the delays weighted by their power is given by the second order moment of the impulse response:

$$D = \left[\frac{\int (t-\mu)^2 h^2(t)dt}{\int h^2(t)dt} \right]^{1/2} \tag{5.11}$$

The temporal integration limits extend from $-\infty$ to $+\infty$. For a given configuration (transmitter, receiver, reflective objects), the impulse response is fixed in time; it is the same for the rms delay spreads.

The normalized delay spread D_N is a dimensionless parameter defined by the ratio of the rms delay spread D and the bit of information duration T:

$$D_N = \frac{D}{T} \tag{5.12}$$

The normalized delay spread constitutes an excellent parameter to determine the power necessary to mitigate multipath.

Let us define function P_{avg} as the average optical transmitted power required to achieve the bit error rate (BER) on the propagation channel characterized by its impulse response $h(t)$ in the presence of a Gaussian white additive noise of power spectral density N_0 using the modulation scheme MS. Then the standardized power required is defined by the following relation [KAHN, 1997, CARRUTHERS, 1997]:

$$\frac{P_{avg}}{P_{OOK}} = \frac{P\big(BER, h(t)/G_0, N_0, MS\big)}{P\big(BER, \delta(t), N_0, OOK\big)} \qquad [5.13]$$

This is the additional power required for the modulation and the channel relative to a non dispersive channel using the ON-OFF Keying (OOK) modulation. As the channels characterized by $h(t)/G_0$ and $\delta(t)$ have the same optical attenuation, only the dispersion effects due to the presence of multipaths are taken into account. The binary error rate is generally taken as equal to 10^{-9}.

5.4. Modeling

5.4.1. *Single reflection propagation model*

In an "indoor" environment, ceiling and walls play an important part in infrared radiation propagation. We present an infrared propagation model in quasi diffuse configuration below, by considering only one reflection of the emitted signal. Named "single reflection propagation model", it will be used to estimate the errors on attenuations introduced by the use of LAMBERT model to represent some surface reflections in a diagram.

This model supposes that only one surface reflects the emitted signal. It enables us to evaluate the impulse response of the quasi diffusive channels. The reflective surface is divided into a large set of small elementary reflective areas called "reflector elements". The spatial and temporal distributions of the emitted signal are evaluated for each of these elements. Each of these elements is then regarded as a point source that emits the collected signal affected according to its reflection coefficient. The reflection pattern of each element is represented by either the Lambert or Phong models described earlier. The received signal is the sum of the signals which arrive at the receiver after reflection on the various elements. Due to the difference in propagation lengths of the various paths, the received signal is dispersed over time. Here, we will only evaluate the effects of the reflection model on the channel propagation losses and we will not consider the delays in signal propagation. The accuracy of the model will increase with the reduction in the size of reflective area of each element [KAHN, 1997, CARRUTHERS, 1997].

Figure 5.8 below illustrates the geometry of the problem allowing us to describe the single reflection propagation model. Let us consider an important surface S made up of areas elements of $\Delta A = \Delta s * \Delta s$ where Δs is the spatial resolution. Let us also consider a transmitting source and a detector, both directed towards the reflection surface. The source emits a power Pt over a half-power angle hpa. The detector consists of a collecting surface A_D and a field-of-view fov.

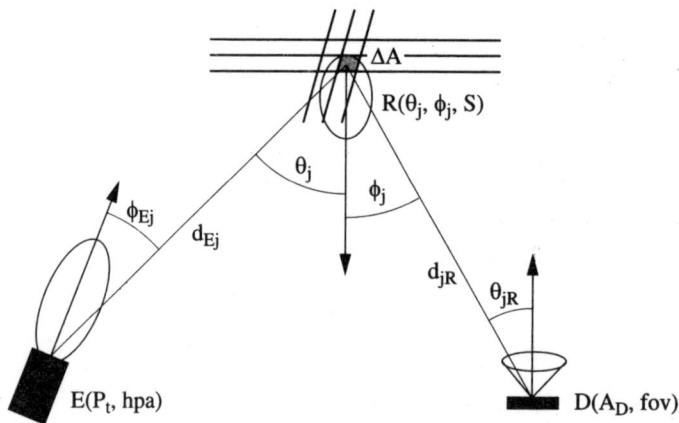

Figure 5.8: *Illustration of the geometry used to describe the single reflection propagation model*

The incident power on a differential element j is given by the relation:

$$P_j(t) = P_t \frac{m+1}{2\pi} \cos^m \left(\phi_{Ej} \right) \cos \left(\theta_j \right) \frac{\Delta A}{d^2_{Ej}}$$ [5.14]

where:

  − ϕ_{Ej} is the angle of emission of the emitted signal,

  − θ_i is the angle of incidence of the signal emitted on reflective surface,

  − d_{Ej} is the distance between the transmitter and the reflector j,

  − m is an integer which controls the directivity of the specular component of the reflection.

The precision of this equation depends on the validity of the assumption: $d_{Ej} \gg \Delta s$.

The power collected at the receiver after being reflected by the element j is given by the following relation:

$$P_{jR}(t) = P_j(t) R(\theta_j, \phi_j, S) \frac{D(\theta, fov)}{d^2_{jR}}$$ [5.15]

where $R(\theta_j, \phi_j, S)$ is a function which represents the reflection pattern of the element j.

This pattern depends on the reflection characteristics of the surface S and is represented either by the Lambert or the Phong model.

The total received power $P_R(t)$ after a reflection on surface S can be calculated by adding all the $P_{jR}(t)$ contributions which are in the field-of-view of the receiver. It is given by the following expression:

$$P_R(t) = \sum_{j=1}^{M} P_t \frac{(m+1) A_D \Delta A}{2\pi d_{Ej}^2 d_{jR}^2} \cos^m(\phi_{Ej}) \cos(\theta_j) R(\theta_j, \phi_j, S) \cos(\theta_{jR})$$ [5.16]

where:
- M is the total number of differential elements which are in field-of-view of the receiver,
- A_D is the collecting surface of the receiver.

It should be noted that this equation is valid for $d_{jR} \gg \sqrt{A_D}$ and that the precision of the result increases with the space resolution.

The required value of Δs will depend on various factors such as:
- the distance between the transmitter and the reflection surface,
- the radiation pattern of the transmitter,
- the distance between the reflection surface and the receiver,
- the field-of-view of the receiver,
- the reflection characteristics of the surface.

5.4.2. Statistical model

5.4.2.1. Free-space loss

Considering Lambertian radiation, neglecting the contribution of the indirect propagation paths and assuming a receiver field-of-view angle of 180°; the line of sight channel loss, in the absence of shadowing, is given approximately by the relation [KAHN, 1997, CARRUTHERS, 1997]:

$$Aff_{free\_space} = -10\log_{10}\left[\frac{A_D}{\pi}\frac{h^2}{\left(h^2+d^2\right)^2}\right]$$ [5.17]

where:

  – h is vertical separation between the transmitter and the receiver,

  – d is their horizontal separation,

  – A_D is the collecting surface of the detector.

In the case of a line of sight configuration, the theoretical values are very close to the experimental values for low values of horizontal separation (about 48 to 66 dB at a horizontal distance of 4.7 m). For higher values of horizontal separation the theoretical curves generally over-estimate the experimental values by about 2 to 3 dB: the effects of the multiple paths are not negligible.

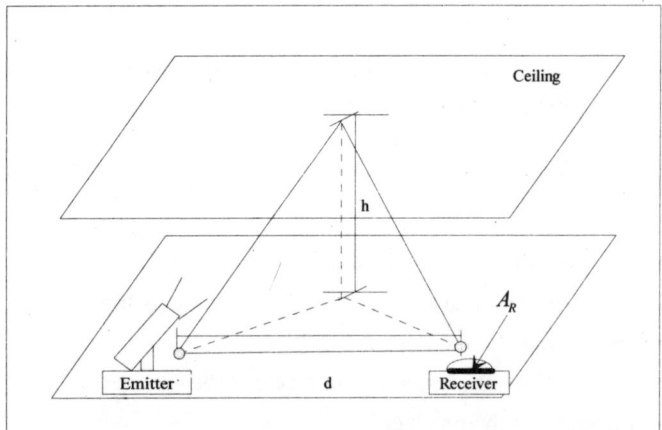

Figure 5.9: *Schematic representation of the link*

In the case of a diffuse configuration the measured loss values are higher than those measured in a line of sight configuration (about 53 dB to 64 dB for a horizontal separation distance of 4.7 m). The theoretical value is evaluated using the equation below:

$$Aff_{diffus} = -10\log_{10}\left[\frac{\rho A_D h_1^2 h_2^2}{\pi^2}\right] * \dots$$

$$\dots \iint_{ceiling} \frac{dxdy}{\left(h_1^2 + x^2 + y^2\right)\left[h_2^2 + \left(x-x_2\right)^2 + \left(y-y_2\right)^2\right]^2}$$

[5.18]

It is supposed here that the initial bound reflection on the ceiling diffuses in a field-of-view angle of 180° with a Lambertian reflector reflectivity ρ. The transmitter and the receiver are located respectively at the coordinates (0,0) and (x_2,y_2) in the horizontal plane (x,y). The parameters h_1 and h_2 are the vertical distances between the transmitter and the ceiling, and the receiver and the ceiling respectively.

5.4.2.2. Impulse response

We present below a statistical model of an infrared indoor diffuse propagation channel based on the estimate of the parameters of the impulse response $h(t)$ (mean delay, RMS delay spread). The Rayleigh and Gamma distributions will allow the representation of the shape of the impulse response $h(t)$ from these values [PEREZ-JIMENEZ, 1997].

For a given position of the transmitter and the receiver, the channel can be considered as stationary because it varies very little in comparison with the binary rate. The reflection model considers all surfaces as Lambertian reflectors independent of the incident signal. All reflections are regarded as additive because the active surface of the receiver is about 10,000 times larger than the wavelength used (see Figure 5.6).

Several channel configurations were tested in an empty square room (10 ×10 × 3 m³). The results were compared for various values of n (from 1 to 255) of the Lambert model. The field-of-view angle varied from 90° to 10°. For each configuration, the two following cases are considered:

– the transmitter and the receiver point towards the centre of the ceiling (see Figure 5.10),

– the transmitter and the receiver point vertically towards the ceiling.

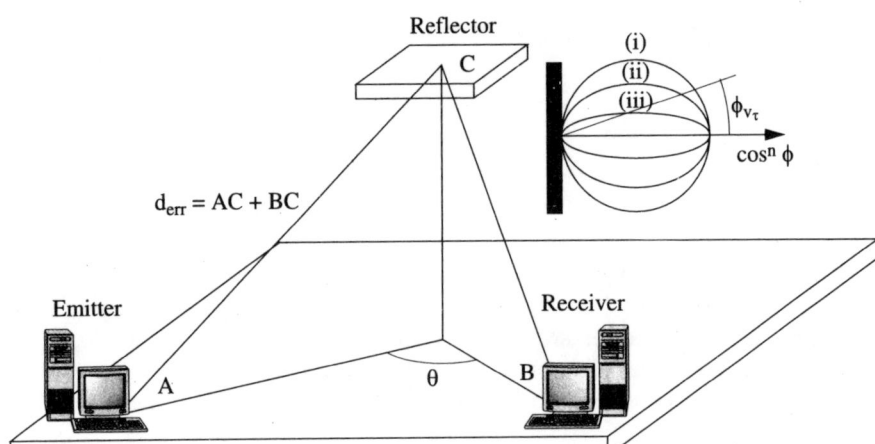

Figure 5.10: *Definition of the variables used to calculate the various parameters*

The distance d_{err}, sum of AC and CB, is the distance traveled by the wave between the transmitter A and the receiver B after reflection on the point C. The angle θ is the angle of transmission between the transmitter and the receiver. The coefficient n is a parameter which controls the shape of the pattern of reflection at the reflection surface; it determines the directivity of the specular component of the reflection. The reflection patterns mentioned on the figure are characterized respectively by (i) $n = 1$, (ii) $n = 3$ and (iii) $n = 50$.

In an infrared link it is important to know both the received total power (gain) and its temporal distribution. The problem is how to obtain the function $h(t)$ characterizing the impulse response of the channel. The process comprises three stages:

– the estimation of the distribution parameters: the RMS delay spread dispersion (τ_{rms}), the mean delay τ_m,

– the adjustment of the impulse response shape (generally by means of Rayleigh or Gamma functions),

– the calculation of the received total power.

The study carried out by Perez-Jimenez *et al.* [PEREZ-JIMENEZ, 1997] showed that τ_m and τ_{rms} depend mainly on the distance d_{err} between the transmitter and the receiver via the reflector point; on the angle of transmission θ between the transmitter and the receiver; and on parameter n characterising the directivity of the specular component of reflective surface ($n= 1$, 3 or 50 (see Figure 5.10)).

The general expressions for τ_{rms} and τ_m are written in the following form:

$$\tau_{rms}(ns) = -0.82n^{0.03} + 0.58n^{-0.11}d_{err} + \left(-0.54 + 0.19d_{err}\right)\cos\left(0.019\theta - 0.32\right) \quad [5.19]$$

$$\tau_m(ns) = -0.46n^{0.28} + 0.33n^{-0.17}d_{err} + K\left(n, d_{err}\right)\cos\left(0.018\theta - 0.012\right) \quad [5.20]$$

where $K(n, d_{err}) = (-0.17n^{0.41} + 0.18n^{0.26}d_{err})$ $\quad$ [5.21]

In the second configuration, where the transmitter and the receiver point vertically both towards the ceiling, the parameters τ_m and τ_{rms} depend mainly on the distance d_{err} between the transmitter and the receiver.

The general expressions for τ_{rms} and τ_m are then written in the following form:

$$\tau_{rms}(ns) = -2.37 + 0.007n + (0.8 - 0.002n)d_{err} \quad [5.22]$$

$$\tau_m(ns) = -3.26n^{-0.04} + (1.12n^{-0.03})d_{err} \quad [5.23]$$

Figures 5.11 and 5.12 below show a comparison between measured and estimated values parameters τ_m (average delay) and τ_{rms} (delay spread).

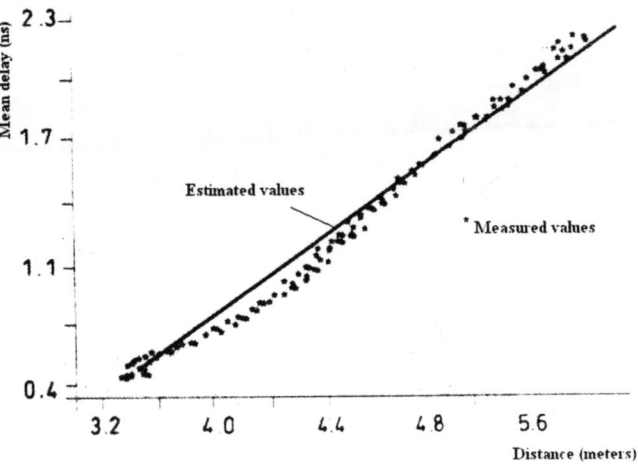

Figure 5.11: *Comparison of the measured and estimated values I_m when the transmitter and the receiver point vertically*

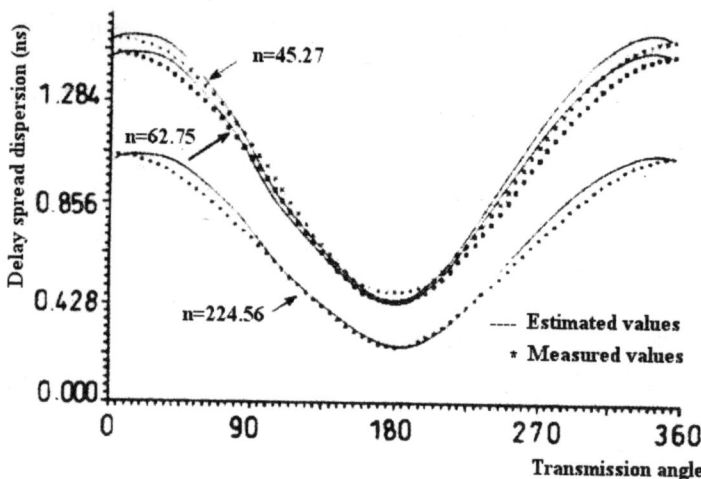

Figure 5.12: *Comparison of the measured and estimated values τ_{rms} of the transmission angle θ for three configurations of n inside a square room of $10 \times 10 \times 3\ m^3$*

The parameters τ_m (mean delay) and τ_{rms} (delay spread) are sufficient to estimate the Multipath Power Penalty (*MPP*) induced by the channel to reach a given Bit Error Rate (BER) [PEREZ-JIMENEZ, 1995]:

$$MPP = 10\log_{10}\left[\frac{Q^{-1}\left(2^M BER_0\right)}{\left(1-\frac{2\tau_{rms}}{T}\right)Q^{-1}\left(BER_0\right)}\right]$$

[5.24]

where:

– BER_0 is the theoretical without Inter Symbol Interference (ISI),

– T is the period of the data,

– M is the length of the Inter Symbol Interference component,

– $Q(x)$ is defined by the following equation:

$$Q(x) = \frac{1}{\sqrt{2\pi}} \int_x^{\infty} e^{-y^2/2} dy$$

Figures 5.13a and 5.13b below show two examples of impulse responses obtained in two different rooms (room A and room B), for various conditions of propagation with or without shadowing (line of sight, diffuse reflection) [KAHN, 1995]:

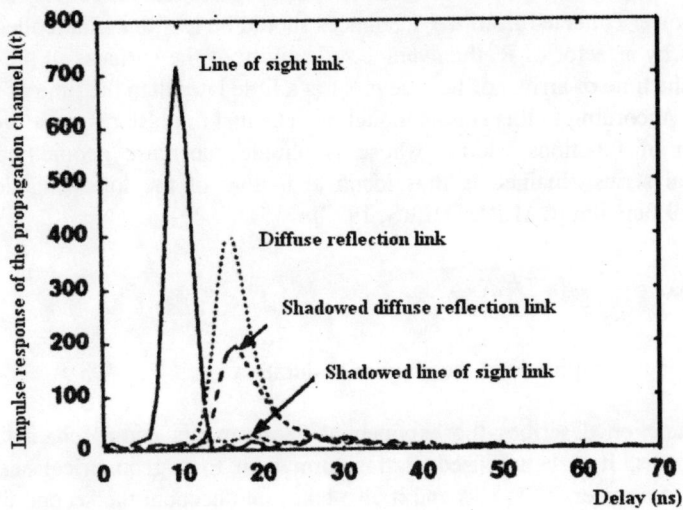

Figure 5.13a: *Example of impulse response of the propagation channel obtained by the reverse transformation of Fourier using a 300 MHz HAMMING window (room A)*

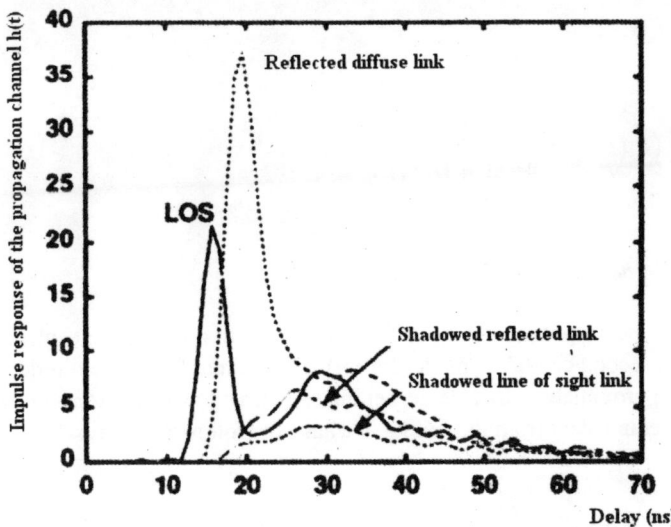

Figure 5.13b: *Example of impulse response of the propagation channel obtained by the reverse transformation of Fourier using a 300 MHz HAMMING window (room B)*

5.4.3. *Exponential decay model*

The light received at the receiver comes either directly or after a certain number of successive reflection paths. As each reflection surface has reflectivity lower than one, the power collected from $n + 1$ bounces should be less than that collected from n bounces by a factor of R, the average reflectivity of the surfaces in the room. In addition, the time of arrival of bounce $n + 1$ is a little later than the time of arrival of bounce n. According to this simple model, the channel impulse response would be a succession of functions "delta" whose amplitudes decrease geometrically. The geometrical series obtained is thus identical to that of the following decreasing exponential function [CARRUTHERS, 1997]:

$$Q(x) = \frac{1}{\sqrt{2\pi}} \int_x^\infty e^{-y^2/2} dy \qquad [5.25]$$

where $u(t)$ is the amplitude of the impulse of duration τ.

This function describes the exponential decay model and is characteristic of multiple paths. It is better used in this form than in a geometrical series form because it is simpler to express and it also takes into account the second dispersive influence on the impulse response, diffuse reflection. The delay spread and the 3 dB cut-off frequency for this model are determined by the following equations:

$$D\big(h_e(t,\tau)\big) = \frac{\tau}{2} \qquad [5.26]$$

$$f_{3dB}\big(h_e(t,\tau)\big) = \frac{1}{4\pi D\big(h_e(t,\tau)\big)} \qquad [5.27]$$

5.4.4. *Ceiling bounce model*

The impulse response due to diffuse reflection from an infinite plane reflector constitutes a second candidate model for $h(t)$. The infinite Lambertian reflecting plane is a good approximation to a large ceiling. Assuming that the transmitter and the receiver are equidistant from the ceiling, we have the following expression for $h(t)$:

$$h(t) = \frac{c\rho AH^4}{\pi} \frac{\frac{ct}{2} u\left(t - \frac{2H}{c}\right)}{\left(\frac{ct}{2}\right)^4 \left(\frac{ct}{2}\right)^4} = \frac{2^7 \rho AH^4}{c^6 \pi t^7} u\left(t - \frac{2H}{c}\right) \qquad [5.28]$$

where:

- ρ is the reflectivity of the plane,
- A is the reception area of the photodiode,
- H is the height of the ceiling above the transmitter and the receiver.

The minimum time required for a signal to travel the path between the transmitter and the receiver via a reflection on the ceiling is equal to $2H/c$. For consistency with the exponential decay model, we eliminate this time by shifting the time origin by $2H/c$. We then obtain the following relation for $h(t)$:

$$h(t) = \frac{\rho A}{3\pi H^2} \frac{6\left(\dfrac{2H}{c}\right)^6}{\left(t + \dfrac{2H}{c}\right)^7} u(t) \qquad [5.29]$$

Let $a = \dfrac{2H}{c}$ and $\dfrac{\rho A}{\left(3\pi H^2\right)} = G_0 = 1$, then we have:

$$h_c(t,a) = \frac{6a^6}{(t+a)^7} u(t) \qquad [5.30]$$

This relation defines the impulse response of the ceiling bounce model for multipath dispersion. The delay spread and the 3 dB cut-off frequency of the model are given by the following relations:

$$D(h_c(t,a)) = \frac{a}{12}\sqrt{\frac{13}{11}} \qquad [5.31]$$

$$f_{3dB}(h_c(t,a)) = \frac{K}{4\pi D(h_c(t,a))} \qquad [5.32]$$

where the factor K is given by the equation:

$$K = \left| \int_0^\infty \frac{\exp(-j.6Ku\sqrt{11/13})}{(1+u)^7} du \right| = \frac{1}{6\sqrt{2}} \approx 0.925 \qquad [5.33]$$

5.5. Additional power required to reach a given bit error rate

5.5.1. *Additional power requirement and delay spread*

We shall next study the relationship between the normalized delay spread D_N and the power required to mitigate multipaths P_{avg}/P_{OOK} for three types of modulation schemes: One-Off Keying (OOK), Pulse Position Modulation (PPM) and Multiple-Subcarrier Modulation (MSM) [CARRUTHERS, 1997, KAHN, 1997].

5.5.2. *One-Off Keying (OOK) modulation*

We consider below the power required for baseband OOK modulation at bit rates of 10, 30, 55 and 100 Mbits/s with three different detection methods: unequalizer, equalizer using a Zero-Forcing Decision Feedback Equalizer (ZF-DFE) and maximum likelihood sequence detection (MLSD).

The transmitter encodes the bit "1" in the form of a rectangular pulse of duration T where $1/T$ is the bit rate. If no equalizer is used, the receiver filter is a five-pole Bessel receiver filter whose 3dB cut-off frequency is equal to $0.6/T$. In the case of using a ZF-DFE equalizer, an identical filter is used but the 3 dB cut-off frequency is equal to $0.45/T$. These various filters perform better than rectangular impulse response filters and their cut-off frequencies were selected in order to provide optimum performances with measured channel responses. For MLSD detection an optimal whitened-matched filter is used.

Figures 5.14 and 5.15 below illustrate the dependence of the additional required power induced by the propagation channel (in presence of multiple paths) to reach the bit rates of 10, 30, 55 and 100 Mbits/s in two cases: (a) without equalization and (b) with maximum likelihood sequence detection. The solid lines show the relationships between required powers and the delay spread for channel impulse responses of the following form:

$$h(t) = \frac{u(t)}{(t+a)^7} \qquad\qquad [5.34]$$

The required powers are relative to those required by OOK modulation in a non distorting channel (absence of multiple paths) having the same optical path loss as the multipath channel.

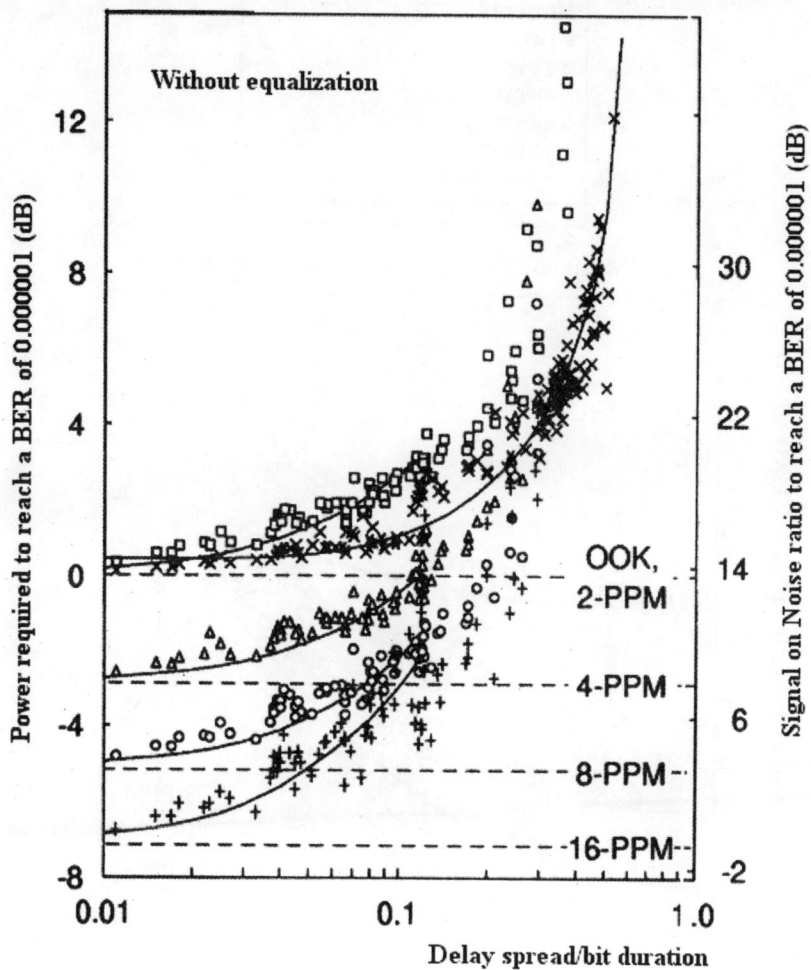

Figure 5.14: *Dependence of the additional required power induced by the propagation channel (presence of multiple paths) to reach the bit rates of 10, 30, 55 and 100 Mbits/s (without equalization)*

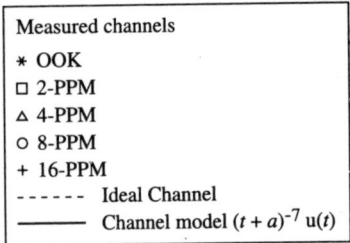

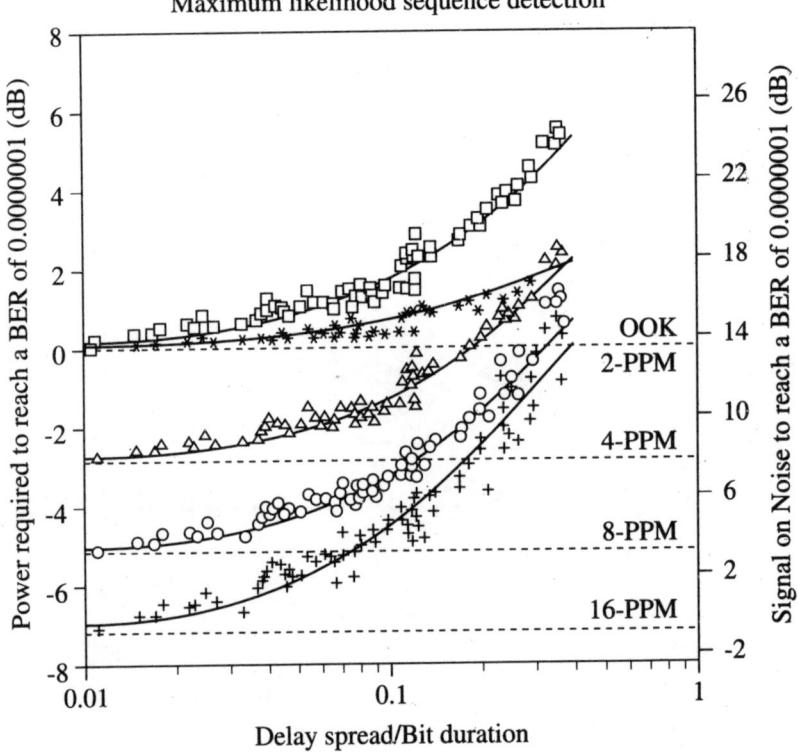

Figure 5.15: *Dependence of the additional required power induced by the propagation channel (presence of multiple paths) to reach the binary rates of 10, 30, 55 and 100 Mbits/s using the maximum likelihood sequence detection (MLSD)*

The result is that, in the "without equalization" case (Figure 5.14), the required power necessary increases exponentially until approximately 13 dB for $D_N = 0.5$. For values of D_N higher than 0.55 the majority of the measured values do not allow the desired performances (TEB = 10^{-9}) to be reached, due to the severity of the Inter Symbol Interference.

In the case of using a maximum likelihood sequence detection (MLSD), the required power necessary increases approximately quadratically with the normalized delay spread up to 2 dB for $D_N = 0.3$ (Figure 5.15).

In the case of using an equalizer ZF-DFE, the required powers are lower about 0.1 dB than those required by using maximum likelihood sequence detection in the same range of normalized delay spread.

5.5.3. *Pulse Position Modulation (PPM)*

Figures 5.14 and 5.15 above also show the relationships between the required powers and the normalized delay spread for various L-position PPM where L is one of 2, 4, 8 or 16 (the ordinate values considered are located on the right hand side of the figures). The emitted impulse and the receiving filter are rectangular. The bit rates considered are equal to 10 and 30 Mbits/s. Two cases, without equalization and with maximum likelihood sequence detection (MLSD), were considered.

5.5.4. *Multiple-Subcarrier Modulation (MSM)*

Figure 5.16 below shows the relation between the required power and D_N for a link using two subcarriers, each one of them using a 4-QAM modulation. The subcarrier centre frequencies are $1/(4T)$ and $3/(4T)$ where $1/(4T)$ is the symbol rate and $1/T$ is the bit rate. The bit rates considered are 30, 55 and 100 Mbits/s. The solid line shows the relation between the required powers and the normalized delay spread for channel impulse responses of the same form as that used above.

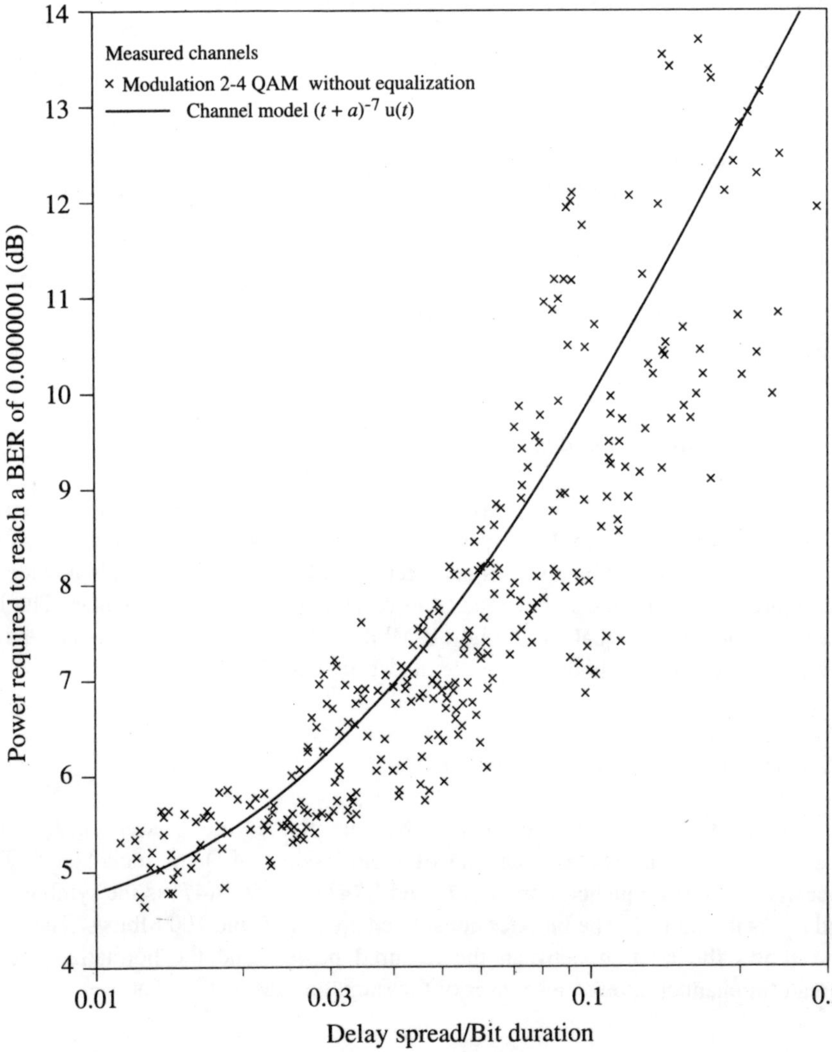

Figure 5.16: *Relation between the required power and D_N for multiple-subcarriers modulation. Two 4-QAM subcarriers are employed*

5.6. Optical noise

In order to design the resident circuit in a confined space, knowledge of the ambient optical noise is essential. We next present an experimental characterization of the optical noises most commonly encountered inside buildings (tungsten filament light, low and high frequency fluorescent light, infrared audio transmitters, remote control of TV systems, sunlight, etc.). We will give, for each, the optical spectrum and the practical means to reduce their influence on the performance of IR links.

The term "ambient noise" is used in a broad sense and no distinction is made between noise, random process, and interference.

5.6.1. *Tungsten filament light*

Tungsten filament lamps are common in offices as well as in the domestic environment. Infrared links between mobiles or microcomputers can be used to print or transfer files between systems in the proximity of light emitted by tungsten filaments.

Figure 5.17 below shows the optical spectrum of a 60 W light source emitted by a tungsten filament. It is very broad and continuous; it shows a maximum towards 1000 nm.

5.6.2. *Low frequency fluorescent light*

Generally installed in or near the ceiling of a room, their influence on the performance of infrared links is of great interest.

The optical spectrum of a low frequency fluorescent tube is represented in Figure 5.17 below. It consists of a continuous part in the visible spectrum and of discrete lines, invisible to the eye, extending up to 1500 nm.

5.6.3. *High frequency fluorescent light*

The use of high frequency fluorescent tubes is currently not very widespread in Europe, unlike in the United States and Japan. They are characterized by the absence of flickering which makes them more comfortable than low frequency fluorescent tubes. They pose severe constraints to free-space infrared links designers and it is thus important to be familiar with their emitted spectrum. Figure 5.17 below shows the spectrum of a ST1 70 HF Philips tube whose frequency of commutation is 28.8 kHz in the wavelength range going from 600 to 1100 nm. This is characterized by a set of lines. Those in the visible spectrum have larger amplitudes.

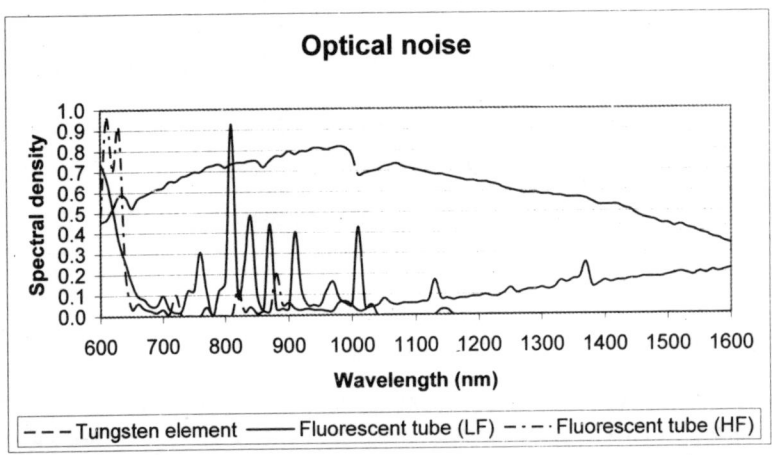

Figure 5.17: *The optical spectrum of a tungsten filament, a low (LF) and high frequency (HF) fluorescent tube*

5.6.4. Infrared audio transmitters

Many models exist on the market. The models analyzed below are the TMR IF310 and TMR IF5 models. The infrared audio modulated transmitting units are positioned in line of sight of a point source of light emitting a set of lines which can disturb infrared links.

Figure 5.18 below shows the spectrum emitted by an infrared audio transmitting station. The Sony LED emits at 870 nm, a typical value for infrared links. The light emitted by emitting LEDs, detected by the photodiodes, interferes with that of the infrared link; the generated significant noise degrades the performance of the link.

5.6.5. Infrared TV remote control devices

This is certainly the most widespread infrared emitting domestic accessory. Its very widespread use justifies the examination of its potential for interference with infrared links. The spectrum emitted by such systems is represented in Figure 5.18 below. Assuming that other infrared links also operate in this wavelength range, interference is possible. In such systems, operating at 970 nm, optical filters to daylight cannot be used because they eliminate the radiation from the infrared beam at the same time.

5.6.6. *Effects of daylight*

The effects of daylight have been studied by many authors [GFELLER, 1979, BARRY, 1991]. Its unfortunate effects on the propagation of infrared beams are well-known. The majority of indoor links do not operate in outdoor environments. The optical broadband and the intensity of daylight saturate the infrared links by generating noise. It also depends on the position of the sun relative to the receiver of the infrared link. The use of optical filters designed to reduce daylight does not bring significant improvement. Figure 5.18 below shows the spectrum emitted by the sun after being daylight filtered. The spectrum varies in intensity and the amplitude of the absorption lines depends on daily conditions. The potential intensity of the solar spectrum between 700 and 870 nm can generate a significant white noise and saturate the majority of infrared links.

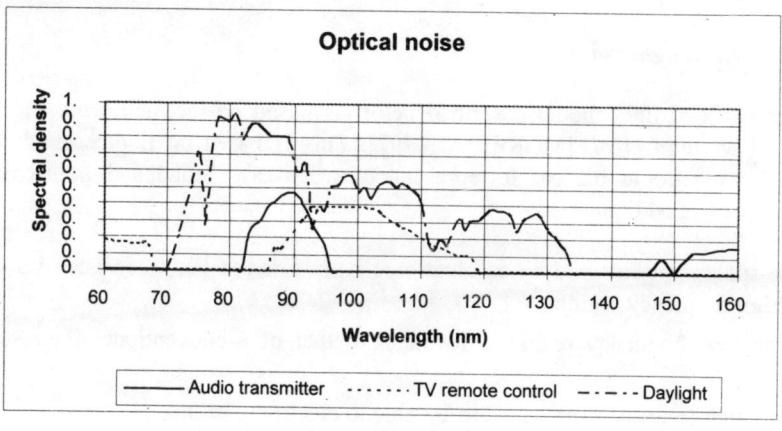

Figure 5.18: *The optical spectrum of an audio transmitter, a TV remote control device and daylight*

Figure 5.19 below shows the spectrum of the three principal sources of noise present inside 'the buildings, namely fluorescent lamps, incandescent lamps and daylight [RAMINEZ-INIGUEZ, 1999].

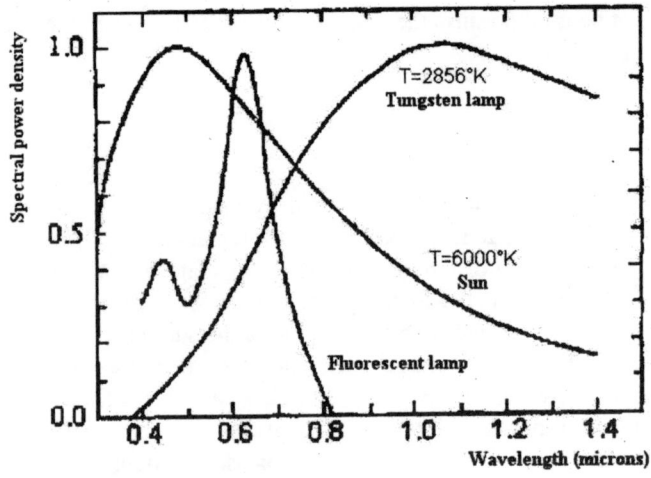

Figure 5.19: *Spectral power density of the three principal sources of light (tungsten lamp, fluorescent lamp, sun)*

5.6.7. Artificial light model

The artificial light model described below is based on measurements and the model of Moreira *et al.* [MOREIRA, 1997]. This is based on fluorescent lamps driven by electronic ballast. It represents a worst-case broadband interference caused by artificial light.

The spectrum generated by an electronic ballast driven lamp consists of a low and a high frequency region:

– the low frequency region is identical to that of a conventional fluorescent lamp,

– the high frequency region is attributable to electronic ballast.

The interference noise at the output of a photodiode can be, expressed by the following mathematical equation [WONG, 2000]:

$$b(t) = RP_m + \frac{RP_m}{A_1} \sum_{i=1}^{i=20} \left[b_i \cos\left(2\pi(100i-50)t+\zeta_i\right) + c_i \cos\left(2\pi(100i)+\varphi_i\right) \right]$$

$$+ \frac{RP_m}{A_2} \left[d_0 \cos\left(2\pi f_h t + \theta_0\right) + \sum_{j=1}^{j=11} d_j \cos\left(2\pi(2jf_h t)+\theta_j\right) \right] \qquad [5.35]$$

where:

- R is the responsivity of the photodiode (A/W),

- P_m is the average optical power of the interfering signal,

- A_1 and A_2 relate the interference amplitude to P_m; their numerical values are equal to 5.9 and 2.1 respectively,

- f_h is the fundamental frequency of the high-frequency component; its numerical value is equal to 37.5 Khz,

- b_i and c_i are the amplitudes of the low-frequency components; they are given by the expressions (36) and (37) below,

- ζ_i and φ_i are phase parameters; their values are given in Table 5-3 below,

- d_j and θ_j are the amplitudes and the phases respectively of the various high frequency components; their values are given in Table 5-4 below.

$$b_i = 10^{(-13.1\ln(100i-50)+27.1)/20}, \quad 1 \le i \le 20 \tag{5.36}$$

$$c_i = 10^{(-20.8\ln(100i)+92.4)/20}, \quad\quad\quad 1 \le i \le 20 \tag{5.37}$$

i	ζ_i	θ_i	i	ζ_i	θ_i
1	4.65	0.00	11	1.26	6.00
2	2.86	0.08	12	1.29	6.17
3	4.43	6.00	13	1.28	5.69
4	3.90	5.31	14	0.63	5.37
5	2.00	2.27	15	6.06	4.00
6	5.98	5.70	16	5.49	3.69
7	2.38	2.07	17	4.45	1.86
8	4.35	3.44	18	3.24	1.38
9	5.87	5.01	19	2.07	5.91
10	0.70	6.01	20	0.87	4.88

Table 5-3: *Typical values of the phase parameters of ζ_i and θ_i*

j	d_j	θ_j	j	d_j	θ_j
0	−22.22	5.09	6	−39.30	3.55
1	0.00	0.00	7	−42.70	4.15
2	−11.50	2.37	8	−46.40	1.64
3	−30.00	5.86	9	−48.10	4.51
5	−33.90	2.94	10	−53.10	3.55
5	−35.30	2.75	11	−54.90	1.78

Table 5-4: *Values of amplitude d_i and phase θ_i parameters for high frequency components*

The three terms of the relation (5.35) represent the photocurrent due to the mean interference, the low frequency interferences and the high frequency interferences respectively. The ratio of the maximum optical power to the mean optical power is approximately equal to $0.6P_m$ for the parameters used in the model.

5.7. Comparison of infrared and radio media and conclusion

As a medium of communication inside the buildings, infrared radiation offers numerous significant advantages when compared with radio-frequency systems.

Transmitters and receivers are able to exchange high-speed information at low costs. The infrared spectral region offers a quasi unlimited bandwidth currently unregulated over the world. The infrared radiation has very similar behavior to visible radiation. It is absorbed by dark objects, diffused and reflected by colored objects, and reflected in a preferential direction by plane surfaces. The two types of light penetrate in glass but cross neither walls, nor other opaque barriers: infrared transmissions are confined to inside the room in which they are generated. This containment of the signal allows all transmissions to easily be secured against eavesdropping and avoids all interference between links operating in neighboring rooms. Infrared wireless local area networks can thus reach a significant capacity; their design is modified owing to the fact that the transmissions in various other rooms do not require any coordination.

Using an Intensity Modulation/Direct Detection (IM/DD) modulation, the very small carrier wavelength and the important detecting area allow good space diversity that negates any multipath fading effects. On the other hand, radioelectric links are subject to important amplitude and phase fluctuations of the received signal.

However, the infrared medium is not without disadvantages: because of the fact that the radiation does not penetrate the walls, communication between one room and another one requires the installation of access points (the equivalent of base stations in mobile radio telecommunications) inter-connected by backbone. In many indoor environments, there is significant ambient optical noise created by the natural light of the sun; incandescent and fluorescent lamps also create significant noise at the input of the receiver. IM/DD modulation is practically the only technique of modulation for indoor applications over short distances. The Signal on Noise Ratio (SNR) at the direct detection receiver is proportional to the square of the received optical power; it only allows for small free-space losses.

The optical losses associated to a diffuse link are higher than those noted in the cases of Line Of Sight or Wide-Line Of Sight (W-LOS). Some experimenters characterize the transmission channel in broadband by using swept frequency networks analyzers [HASHEMI, 1994, KHAN, 1995]. In a typical configuration the link losses are about 120 to 130 dB with a mean delay equal to 12 ns. The design of the receiver must be more sophisticated owing to the fact that it must have the dynamics to be able to follow important signal intensity variations.

The characteristics of the diffuse channel are dependent on transmitter and receiver alignment on the one hand (existence of the direct path, shadowing); and on the environment inside the room in terms of furniture and building materials on the other hand.

Systems currently marketed are based on diffuse links operating at date rates going from 1 to 4 Mbits/s. Standard IEEE802.11 includes a definition of the infrared diffuse physical link to complement that of the radio link. Demonstrations in laboratories of operative systems at 50 Mbits/s and the feasibility studies of systems at 100 Mbits/s by using adaptive equalizers are mentioned in the literature [AUDEH, 1995, MARSH, 1994]. However, the design of the optical concentrator and the filter still constitutes a challenge in terms of volume.

Chapter 6

Optical Communication

6.1. A reminder about digitization

The phenomena that surround us are, for the most part, continuous phenomena. They are quantifiable; they pass from one value to another without interruption. When you wish to restore or reproduce these values, it is necessary, first, to make a recording on a physical medium. When the physical medium can take continuous values, this is called analogue recording. For example a video cassette, an audio cassette or a disc vinyl are analogue media; but their frequent or prolonged use involves eventual deterioration of the medium and thus of the recording.

A solution to this problem is to proceed with the digitization of the signal to be reproduced, i.e. "to cut out" these continuous phenomena in the smallest "pieces" possible, in order to "transform" the continuous signal into a succession of "bricks" more or less the same "height". The example in Figure 6.1 below represents the principle of digitization.

Digitization consists in "replacing" the analogue signal (the black curve), by "a brick" succession in which the height (i.e. the amplitude) is digitized (the black histogram). The transformation of an analogue signal into a numerical, or digital, signal is called digitization or sampling. Sampling consists of periodically taking samples of a signal. The numerical signal quality will depend on two factors:

– the sampling frequency (called sampling rate): the larger the sampling frequency (i.e. the samples are picked up at smaller intervals of time), the closer the numerical signal will be to the original one,

– the number of bits onto which we encode the values (called resolution), i.e. the number of different values that a sample can take. The larger this is, the better the quality.

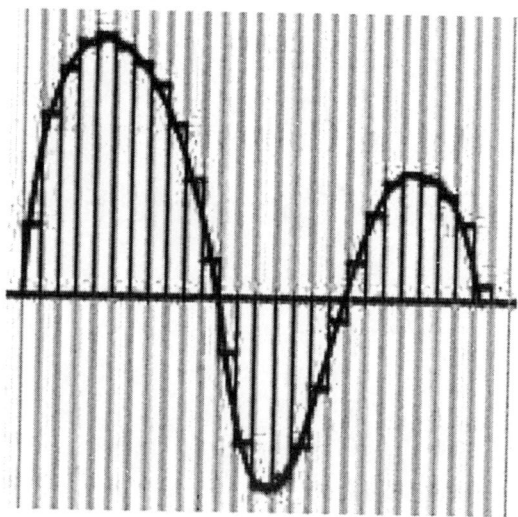

Figure 6.1: *Example of digitization*

An information is known as numerical, or digital, when it "is translated" into the form of basic units of information (bits) having only two values: 0 or 1; which correspond to the passage or the cut of the electrical current or of a light beam, while simplifying it, thus forming a language readable by a computer.

A succession of six 0 and 1 allows us to represent 64 different combinations ($2^6 = 64$); a succession of 12 bits offers 1096 different combinations, etc.

The power of this exponential function is that it allows increasingly complex information to be digitized; texts, images, sounds or video.

With respect to the concept of data transfer, the number of these basic units of information (bits) transmitted per second gives the rate of transfer of this numerical information. Thus, the term kilobits (kbits/s or Kbps) means the transfer of thousands of bits a second, megabits (Mbits/s or Mbps) for millions of bits a second and gigabits (Gbits/s or Gbps) for billions of bits a second.

6.2. Examples of laser applications outside optical communications

The principal characteristic of a laser beam is the propagation of a perfectly straight and not very divergent light beam.

It is thanks to this property that a laser was used at the time of the construction of the Montparnasse tower, which was used to calculate the Earth–Moon distance (the laser beam is reflected on a given surface and the speed of the light being known, it is possible, by measuring the time the laser beam takes to make the return journey, to calculate the distance separating the laser source from an obstacle).

Other uses, covering a significant number of branches of industry, are listed below.

6.2.1. *BTP sector*

– Alignment lasers used for public works, car-manufacturing, public works machines guidance:

- power of the lasers used: approximately 10 mW,

- type of beam: continuous,

- Note – use of small lasers (for instance, helium-neon lasers).

– Cleaning and preparation of surfaces, for instance, scouring historic buildings with lasers:

- power of the lasers used: power peaks of about 100 W (for powers from 10 to 20 W for YAG lasers, for example),

- type of beam: impulse (very short impulses: of a few tens to a few hundreds of nanoseconds),

- Note – the use of YAG lasers. This process allows us to completely eliminate, in a selective way, surface layers covering various materials without damaging the layers below, by concentrating the laser beam on the zones to be removed.

6.2.2. *Industrial sector*

– Metal welding:

- power of the lasers used: according to the thickness of materials, a few tens of watts to 50 kW,

- type of beam: continuous or impulse,

- Note – generally, the use of YAG lasers (100 W to 2 kW) or CO_2 lasers (100 W to 50 kW).

– Cutting materials such as wood, plastic, glass or metals:

- power of the lasers used: 1 to 3 kW,

- type of beam: continuous or impulse.

6.2.3. *Medical sector*

– External surgery: care of the eye:

- power of the lasers used: excimer laser (argon/fluorine, emission at 193 nm),

- type of beam: continuous or impulse,

- Note – it performs a very fine and deep corneal incision allowing shaping of the cornea.

– Surgery: care of the eye:

- power of the lasers used: laser with diode, emission in infrared (810 nm),

- type of beam: continuous or impulse,

- Note – treatment of retinal neovessel and use as endophotocoagulator during intraocular operations.

– Surgery:

- power of the lasers used: Nd-YAG Laser,

- type of beam: continuous or impulse,

- Note – the laser acts as a scalpel.

6.2.4. *General public sector*

– Reading compact disks – reading barcodes:

- power of the lasers used: a few mW,

- type of beam: continuous,

- Note – the diode lasers are integrated into the apparatus like ordinary electronic components.

– Discos, laser shows:

- power of the lasers used: a few Watts,

- type of beam: continuous,

- Note – use of powerful lasers and mainly helium-neon or argon lasers.

– Computer interface, PDA or mobile:

- power of the lasers used: a few milliwatts in the infrared field,

- type of beam: continuous,

- Note – used for numerical file transfer.

6.3. Inter-satellite or Earth–satellite optical communications

6.3.1. *Earth–satellite optical communications*

Free-space optical transmission is not a recent innovation; this technology has already been used for about 30 years, both in the prototype state and for very specific applications, and mainly within a military framework.

The first known project was initiated in the seventies. It had code number "405B". It was, at the time, about "the study of space communication system by laser" within the "Weapons System" project framework, entirely financed by the US Air Force and the Pentagon. NASA, the Goddard Space Flight Center and the Jet Propulsion laboratory participated in this project.

During this same period, and again financed by the American military; telecommunication links between planes were tested using CO_2 lasers as well as Neodymium-YAG (Nd-YAG) lasers. Then, as a logical extension, these experiments were extended to other forms of communication such as ground-to-ground laser, plane-to-plane, and, *a priori*, satellite-to-plane and satellite-to-submarine.

In the 1990s, civil studies were launched in connection with prototypes for inter satellite communications; by using GaAlAs diode lasers, with the aim of transmitting flows of 650 Mbps at a distance of 21000 km. More recently, extensions were studied for communications between the Earth and the modules on mission over Mars (NASA's Space Exploration Initiative – SEI) with the Lewis Research Center laboratory. The aim was to use Nd-YAG, GaAs lasers, and others, to obtain data rates from 10 to 100 Mbps.

6.3.2. *Inter-satellite optical communications*

By the beginning of the year 2000, the first laser transmission and high communication rate between two satellites had taken place. In November 2001, there was a successful link between the French satellite Spot-4 and the European satellite Artemis separated from each other by several tens of thousands of kilometers. This European success was the result of a European project – SILEX Project (Semiconductor Inter-satellite Link Experiment) initiated at the beginning of the nineties, whose partners were Matra Espace, Astrium, SDL, the European Space Agency (ESA) and CNES (Centre national d'études spatiales).

The system consists of terminals equipped with small 25 cm diameter telescopes allowing the control and the pointing of the laser beams that they produce. It uses lasers emitting in the near infrared region and transmitting at a rate of 50 Mbps. Two twenty-minute tests were carried out and, in addition to the absence of any misalignment problem, the transfer of the numerical data was carried out with practically no loss of information: there was roughly 1 erroneous bit for 1 billion transmitted bits (10^{-9} TEB).

The latest projects concerning Earth−satellite communications are directed towards a 2000 kilometer distance laser communication at 1 Gbps.

Figure 6.2: *Satellite-to-satellite laser communication: link between Spot and Artemis using Silex laser system (from ESA)*

6.4. Free-space optical communications

6.4.1. *Introduction – operating principles*

The first direct line of sight laser communication tests were carried out at the beginning of the seventies at Lannion between two buildings of the Telecommunications Studies National Centre (CNET, currently called FTR&D). However, this was only a laboratory experiment, far from the necessary requirements for a good transmission between customers or subscribers.

In the field of atmospheric optical communications or free-space optical links (FSO), the first civil application products made their appearance at the beginning of the 1980s. First systems were tested. However, in spite of unquestionable progress in transmitter and detector technologies, transmission quality and the availability of communication links did not yet answer the expectations of telecommunications operators, and this was mainly for the following reasons:
 − the unreliability of the equipment and imprecise implementation conditions,
 − other technological solutions presented a better ratio,
 − lack of motivation to improve them,
 − satisfactory radio-relay systems links, no license for the use of the spectrum,
 − no competition incentive to find solutions of less expensive point-to-point links,
 − explosion in optical fiber performance.

With the end of the 1990s, a new wave of products was proposed, mainly of European and American origin. The telecommunications market became more competitive; the technical solutions of free-space optical communication were re-examined to determine for which digital communication market segments this technology could be proposed (CNET, 1993 and 1995).

Basically, the principle of laser link transmission is based on the transmission of a divergent beam, in line of sight. In the example given in Figure 6.3, the laser equipment is placed at the top edge of each building and pointed one towards the other. The equipment of site X sends digital information towards site Y by the intermediary of a modulated laser beam. This beam is deliberately slightly divergent in order to reduce the problems of misalignment; then a part of the front wave is collected on the reception diode of the equipment B placed on site Y. This operation is also carried out in opposite direction, from Y towards X.

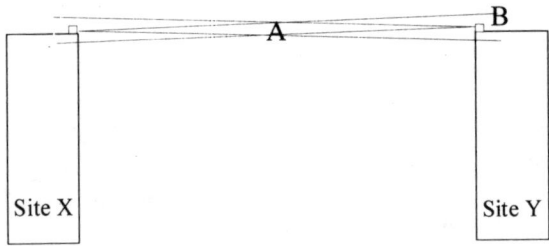

Figure 6.3: *Diagram of a point-to-point Free-Space Optical link; between two sites, X and Y*

More precisely, the equipment uses the modulation of a laser beam to exchange binary data in two directions (Full-Duplex) by the intermediary of a couple emitter/receptor (Laser diode/PIN diode) at each end. It is a point-to-point, bilateral link and in line of sight.

Each piece of equipment consists of several modules:

− for transmission:

    - the connection interface: electric or optic to send and receive numerical data,

    - the electric/optic conversion module (in the event of optical interface),

    - the filtering and amplification of the digital electric signal,

    - the optical transmission module containing the laser.

− for reception:

    - the optical reception module containing the diode,

    - the filtering and amplification of the digital electric signal,

    - the optic/electric conversion module (in the event of optical interface),

    - the connection interface: electric or optic to send and receive the numerical data.

Management software is sometimes provided with the equipment. This software allows the link to be parameterized and qualitative and quantitative information about the different modules to be obtained.

According to the manufacturers, additional functions can be implemented, such as:

− a system of pointing control (tracking),

− a radio assistance link, with a limited rate, in the event of a laser link interruption.

An example of the structure of FSO equipment is presented below (Figure 6.4).

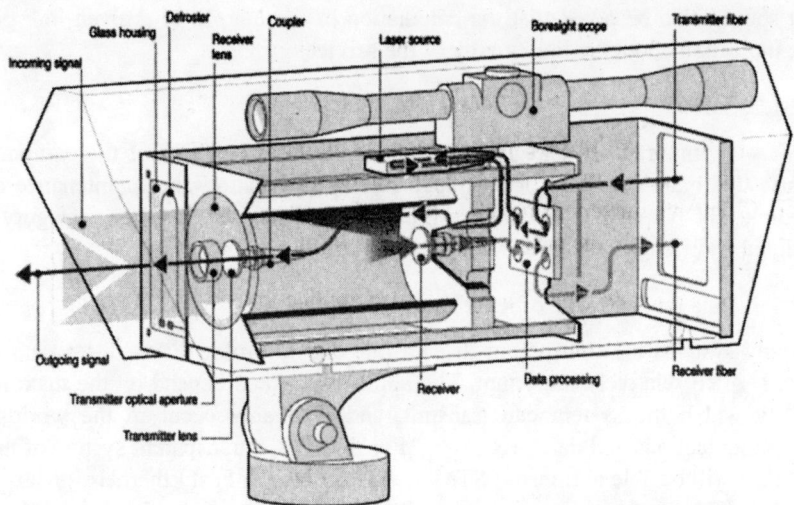

Figure 6.4: *Example of equipment for Free-Space Optical link (according to Optical Access, San Diego USA). We note a strong analogy in the constitution of this equipment with Mangin's optical telegraph, shown in Figure 1.6, proposed more than 120 years ago*

6.4.2. Characteristics

To avoid a too detailed description of the various products, a presentation of the general parameters is given below.

We will restrict ourselves to the products which have the following properties:

– laser technology whatever the wavelength, power and category,

– a numerical data transport function,

– a point-to-point connection or system, in line of sight,

– transmission rate from 0.5 Mbps to 2.5 Gbps or more.

6.4.2.1. Principal parameters

The principal parameters which must be taken into account in the definition of optical links are range, safety and data rate.

6.4.2.1.1. Range

This varies according to the equipment from a few tens of meters to several kilometers. Certain manufacturers give a maximum range, others specify the typical range for various weather conditions, and others propose a "recommended" range, integrating a margin around a maximum value. These values must be taken as orders of magnitude, and not as absolute values.

It should also be noted that the calculation of the margin of a given link gives more indication about the link quality of the service.

6.4.2.1.2. Safety

A factor important to take into account is the laser category of the equipment, because this makes it more or less easy for the installation and maintenance of a laser link. The parameters to be taken into account when defining laser category are the signal wavelength, the power and the beam form.

6.4.2.1.3. Data rate and type of recommended application

Many systems are transparent to data rate and to protocol this, for a data rate range, is often relatively important. The applications then depend on the maximum capacity which the system can transmit, and invariably occur in the worlds of telecommunications and data processing. For instance, a transparent system of up to 200 Mbps will be able to transmit STM-1, ATM, FDDI or Fast Ethernet signals.

Other products are specified for a data rate, an interface and thus a given use, for example the E1 (2.048 Mbps), or Ethernet data (10 Mbps).

6.4.2.2. *Secondary parameters.*

Other parameters should also to be taken into account for the choice of a system such as:

– the wavelength at which the optical link operates: this parameter influences the link margin; and by consequence, the quality of service,

– the type and number of optical transmitters: this also influences the link margin,

– the alignment control: potentially to offer better protection against shock and vibration,

– additional safety components: an automatic beam cut-off in the event of detection, for example, of a security key to access the equipment,

– a simple process of implementation and maintenance,

– simple, convivial supervision software allowing the management of two (or more) elements of links from only one site,

– and, obviously, the cost of the system.

6.4.2.3. *Examples of installed systems*

More than one hundred systems were listed, in 2004, so it is difficult to represent these products by a photograph. For more information, the reader should refer to the manufacturers sites indicated below (Table 6-1).

Names	City	Country	Website
Actipole	Bouscat	France	www.laser-com.com
Advalase Corporation	Los Alamos	USA	www.advalase.com
Air Fiber	San Diego	USA	www.airfiber.com
AirLinx	Boston	USA	www.airlinx.com
Alcatel	Paris	France	www.alcatel.com
Aoptix	Campbell	USA	www.aoptix.com
CableFree	Mampton Hill	England	www.cablefree.co.uk
Canon	Tokyo	Japan	www.canon-europe.com
Communication By Light	Munster	Germany	www.cbl.de
Dominion Lasercom	Bryan	USA	www.dominioncom.com
Efkon	Graz	Austria	www.efkon.com
Fsona	Richmond	Canada	www.fsona.com
GoC	Dreieich	Germany	www.goc.de
Infrared Acropolis	Moscow	Russia	infrared.acropolis.ru
IRLan	Yokneam	Israel	www.irlan.co.il
Katharsis	St Petersbourg	Russia	www.infrared.ru
Lase	Wesel	Germany	www.lase.de
LaserBit Communications	Budapest	Hungary	www.laserbitcommunications.com
LightPointe	San Diego	USA	www.lightpointe.com
LSA	Exton	USA	www.lsainc.com
MRV	Yokneam	Israel	www.mrv.com
Omnilux	Pasadena	USA	www.omnilux.net
Optel	Hamburg	Germany	www.optel.de
PAV Data Systems	Cumbria	England	www.pavdata.com
Plaintree Systems	Arnprior	Canada	www.plaintree.com
RedLine	Kyalami	South Africa	www.redlinesa.com
Silcom	Mississauga	Canada	www.silcomtech.com
Sunflower	Seoul	Korea	www.telsonic.co.kr
TeraBeam Networks	Redmond	USA	www.terabeam.com

Table 6-1: *Some Free-Space Optic manufacturers*

6.4.3. *Propagation times*

Another interesting characteristic of FSO equipment is its speed of transmitting digital data; for instance, this could allow the router for a LAN link to be cast off.

Indeed, the majority of FSO equipment is transparent to the transmitted protocol. Generally, no treatment is carried out on the content or the nature of the data, which offers relatively short propagation times.

The parameters to be taken into account in the calculation of the propagation time of a signal are:

– electronic processing time of the FSO equipment (both Emission and Reception),

– the propagation time of light in the atmosphere between equipment A and B (the propagation time of light in the atmosphere is about 3.10^{-9} s/m).

The total propagation time is the sum of these two parameters.

Example of application: a 500 meter link at 155 Mbps using manufactured FSO equipment:

– the electronic processing time of this FSO equipment is $3.10^{-7} \times 2 = 6.10^{-7}$ s,

– the propagation time of light in the atmosphere between equipment A and B: for 500 m the propagation time is $1.5.10^{-6}$ s,

– the total propagation time is: $6.10^{-7} + 1.5.10^{-6} = 2.1 \times 10^{-6} \Leftrightarrow 2.11$ μs.

6.4.4. *Implementation recommendations*

In general, FSO equipment is set up in a similar way to a radio-relay system:

– installation on a high point (such as a building, pylon or water tower),

– in line of sight, with no obstacles in the present or future trajectory,

– installation time lower than one day for a link.

However, due to the technology used and in addition to the safety requirements related to the laser class equipment, some elements have to be taken into account at the time of the installation.

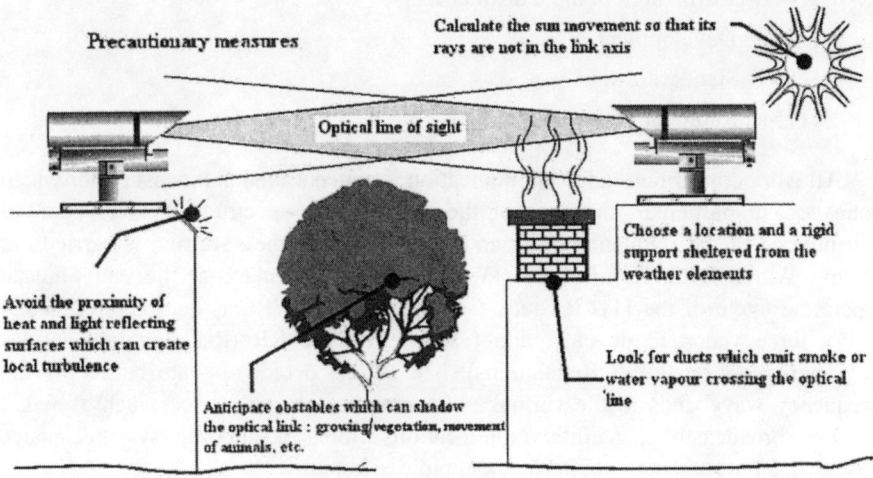

Figure 6.5: *Implementation of Free-Space Optic telecommunication equipment: potential problems to avoid at the time of choice of the place (source: Actipole)*

Very precise alignment is necessary, given the characteristics of the equipment (i.e. low divergence of the laser beam). The coupling of the optical link is characterized by the alignment of the transmitter and the receiver. These can be disturbed following mechanical vibrations. The fitter of the communication system must:

– fix the materials on a rigid support or a load-bearing wall so that it is subjected to less possible vibration or shock (for instance, away from the edges of walls and the sides of walls),

– avoid the direct alignment of optics with the rays of the sun,

– avoid the proximity of elements that can cause atmospheric turbulence (such as chimneys and reflective surfaces).

6.4.5. *Legislation*

6.4.5.1. *The organization of regulation activities in radio communications*

The ITU, whose headquarters are in Geneva (Switzerland), is an international organization within the United Nations System where governments and the private sector coordinate global telecommunications networks and services.

The ITU is structured in three distinct sectors:

- ITU-D: Development,
- ITU-T: Standardization,
- ITU-R: Radiocommunication.

All work concerning radiocommunication is concentrated in the last sector which manages, in particular, the uses of the radioelectric spectrum. This revision of attributions of the frequency plans and the division of the spectrum is carried out during World Radio Conferences (WRC) which take place at the end of each operating cycle of the ITU-R Study Commissions. A WRC is held approximately every three years, to develop, adopt and revise the RR (Radiocommunications Regimentation or Radio Regulations). The WRC decide the attribution of the frequency wavebands to the various radiocommunications services such as fixed, mobile, broadcasting, satellite, radiolocation, global positioning systems, space research, environmental monitoring, and radio astronomy.

Each ITU country must conform to these frequency wavebands and to the conditions of division fixed by the ITU. Attributions and access and sharing conditions are described in the Radiocommunications Regulation. The next WRC will be held from October 8th to November 8th 2007, in Geneva.

At the European level, the CEPT (European Conference of the Post and Telecommunication) is the reference organization. The CEPT gathers 44 countries, including all the EU countries, at the permanent office, the ERO (European Radiocommunications Office), which is based in Copenhagen (Denmark). Based on the Decisions and Recommendations of its Electronic Communications Committee (ECC), CEPT decides, within the framework of the frequency waveband attributions fixed by the ITU-R, the particular conditions which prevail for the use of these attributions: frequency waveband reservation for particular systems (GSM, DECT, etc.), fixations of the cables for fixed service in the various frequency bands which are allocated to them by the ITU-R, technical operating and regulation conditions. These decisions or recommendations aim at harmonizing the uses of frequencies by the various European countries, with the objective of facilitating the development of a European market, as well as solving the problems of coordination at the borders. The various European countries must conform to CEPT's decisions while the recommendations aim at harmonizing the use of the spectrum without being constraining.

The acceleration of the rhythm of WRC, and the need to reach a consensus within the framework of these conferences, led European countries to cooperate in carrying out the WRC preparation together. The role played by CEPT within this framework is fundamental; up to now the European Commission has only been a spectator. The CPG (Conference Preparatory Group), working group from the ECC (Electronic Communications Committee) coordinates this effort and prepares the common European positions for the WRC.

Lastly, in France, the telecommunications regulation law of July 26th 1996 created two new authorities, the Authority of Regulation of Telecommunication (ART) and the National Agency of Frequencies (ANFR). The ANFR (National Agency of the Frequencies) has the role of ensuring the schedule, management and control of the use, including private, of the public radio domain. Based at Maison-Alfort, the agency manages the spectrum globally, together with the frequency waveband attributions (scientific communities, weather service, operators of telecommunication, military, radio sea service, etc.).

ANFR proposes the allocations of the frequency wavebands between the SCA, the ART, and the French administration which are approved by Prime Ministerial decree. The last edition of the "national table of the frequency bands allocation" was published by the Prime Minister's decree on March 6th 2001 (J.O. of March 8th 2001).

6.4.5.2. *Regulation of FSO equipment*

If we go back to Article 1.5 of the Radiocommunications Regulation, this defines radio waves or hertzian waves: electromagnetic waves whose frequency is currently lower than 3000 GHz and which propagate in space without artificial guide.

It appears that the wavelengths used by FSO equipment are not currently covered by the provisions of the Radiocommunication Regulation, which are limited to frequencies lower than 3000 GHz. Indeed, FSO equipment normally functions at frequencies located between 150 and 500 THz.

This is the reason why no legislation or management and attribution of this part of spectrum exist. One of the direct consequences of this peculiarity is the absence of tax or expenses related to license attribution.

It should however be noted that at the 2002 ITU Plenipotentiary Conference, which is the highest body of the International Telecommunication Union (ITU), noting that radio communication techniques showed that it was possible to use electromagnetic waves in space, without artificial guide above 3000 GHz, a new Resolution on the use of the spectrum above 3000 GHz was adopted (Resolution 118 (Marrakech, 2002)): see text below.

This Resolution:

– requests the Radiocommunication Assembly (which was held one week before the 2003 WRC) to "study, within the framework of its work program, if it is possible and if it is necessary to include frequency bands higher than 3000 GHz in the Radiocommunications Regulation",

– authorizes the following WRC to include in their agenda the points relating to the frequencies higher than 3000 GHz and then to introduce, if required, the provisions concerning these frequencies into the RR.

Note – entry into force of such new regulation would depend on consequential changes to no. 1005 of the Annex to the Convention at the next plenipotentiary conference.

RESOLUTION 118 (Marrakech, 2002)

Use of spectrum at frequencies above 3000 GHz

The Plenipotentiary Conference of the International Telecommunication Union (Marrakech, 2002):

– considering:

a) that no. 78 of the ITU Constitution and no. 1005 of the Annex of the ITU Convention allow study groups of the Radiocommunications Sector (ITU-R) to study questions and adopt recommendations dealing with frequency bands without limit in frequency;

b) that studies are being carried out within ITU-R study groups that consider technology operating above 3000 GHz;

c) that the frequency that can be regulated in the Radio Regulations are limited to below 3000 GHz by the definition of "radiocommunications" in no. 1005 of the Annex to the Convention;

d) that radiocommunications technology has demonstrated the ability to use electromagnetic waves in space without artificial guide above 3000 GHz, and that some Member States are of the opinion that the limit of 3000 GHz should be removed to allow competent world radiocommunications conferences to introduce, if needed, provisions in the Radio Regulations;

e) that frequency bands above 300 GHz have been used for a long time, especially in the infrared and visible bands, by systems/applications regulated by national and non-ITU provisions, and that some Member States are of the opinion that the relationship between those provisions and ITU provisions should be thoroughly considered before changing the definition contained in the Convention,

– invites the Radiocommunications Assembly to include, in its program of work, studies of the possibility and relevance of including in the Radio Regulations frequency bands above 3000 GHz,

– instructs the Director of the Radiocommunications Bureau to report to World Radiocommunications Conferences on the progress of ITU-R studies concerning the use of frequencies above 3000 GHz,

> – resolves that World Radiocommunications Conferences can include in agenda for future conferences, items relevant to spectrum regulation of frequency above 3000 GHz and take any appropriate measures, including revision of the relevant parts of the Radio Regulations[1],
> – urges Member States to continue participating in the work taking place in ITU-R on the use of spectrum above 3000 GHz.

6.4.6. *Concept of quality of service and availability*

One of the important elements to know in free-space optical transmissions is the margin of the laser link. In fact, following the example of radio operator equipment or radio-relay systems, it is of primary importance to know the margin of a given link. When a link is installed, mathematical models allow the availability of the link for one year, or for the most unfavorable month, to be calculated.

The first step consists of calculating the link margin. This element allows the capacity of the laser equipment to transmit numerical data in spite of the variations of the weather conditions to be known.

To use the prediction models, the necessary parameters of the equipment are:
– the emitted power,
– the sensitivity of the receiver,
– the capture area of the receiver,
– the divergence of the emitted beam.

From these data, we can work out the value of the link geometrical attenuation, then the link margin and thus its availability.

6.4.6.1. *Geometrical attenuation concept*

The beam emitted by the transmitter being divergent (Figure 6.6), the receiving cell will collect only a fraction of the energy emitted [6.1]:

$$Aff_{geometric} = \frac{S_d}{S_{capture}} = \frac{\frac{\pi}{4}(d\theta)^2}{S_{capture}} \tag{6.1}$$

---

1 Entry into force of such new regulation would depend on consequential changes to no. 1005 of the Annex to the Convention at the next plenipotentiary conference.

where:

    – θ is the divergence of the beam (for example, 1.3 mrad),

    – d is the transmitter – receiver distance,

    – $S_{capture}$ is the capture area of the receiver (for example, 0.005, 0.025 m^2),

    – S_d is the area of the beam at distance D.

In dB, the attenuation is given by: $Aff_{dB} = 10 \log_{10}(Aff)$ [6.2]

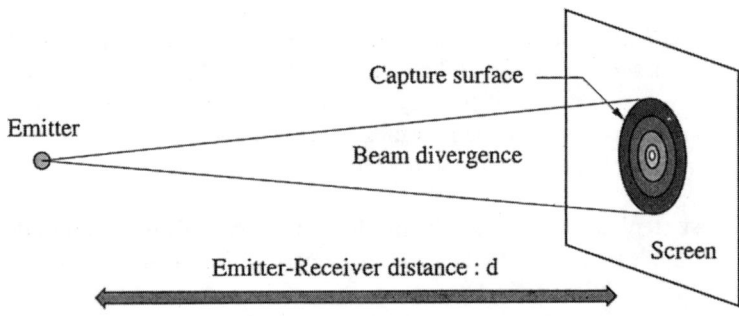

Figure 6.6: *Diagram to show the propagation of a laser beam. The transmitter provides a point source, this emitted beam widens out with the propagation distance, so the effective surface of the detector must be adapted to the section of the beam which is dependent on the transmitter – receiver distance D*

The geometrical attenuation is thus a function of the beam divergence, the distance and the receiver capture area. Table 6-2 below gives some examples of geometrical attenuation values.

Divergence (mrad)	Distance (km)	Geometrical attenuation (dB)
1	5	29
1	8	33
3	2	37

Table 6-2: *Values of geometrical attenuation transmitter – receiver links for various distances*

The graph below (Figure 6.7) gives the variation of geometrical attenuation with distance for various values of beam divergence (1, 2 and 3 mrad), the capture area is constant and equal to 0.025 m^2.

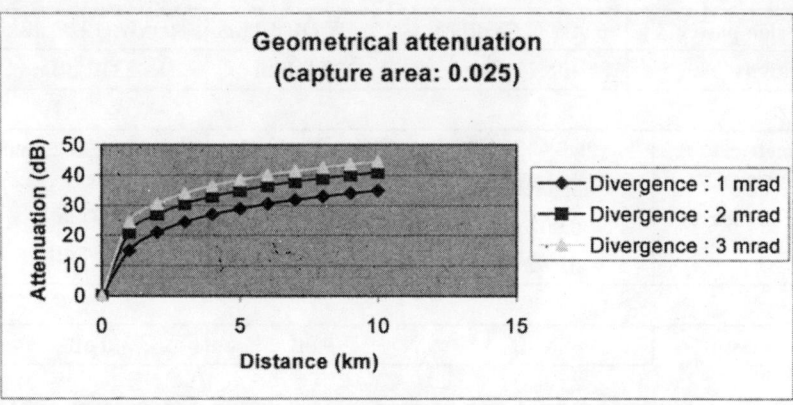

Figure 6.7: *Variation of geometrical attenuation with distance for various values of beam divergence*

6.4.6.2. *Concept of the link margin*

After calculation of the geometrical attenuation of a laser link, the clear margin is determined as follows.

Emission power (*Pe*), receiver sensitivity (*Sr*), data given by manufacturers, allow the calculation of the link margin using the following formula [6.3]:

$$M_{link} = Pe + |Sr| - Aff_{Geo(dB)} - Aff_{Atm(dB)} - P_{Syst(dB)} \qquad [6.3]$$

where:

– M_{link} is the link margin (dB),

– *Pe* is the power of the emission signal (dBm),

– *Sr* is the receiver sensitivity (dBm),

– $Aff_{Geo(dB)}$ is the geometrical attenuation of the link,

– $P_{syst(dB)}$ is the equipment loss of the equipment (dB) given by the manufacturer (possibly multiplied by 2).

In Table 6-3, we demonstrate the calculation of the link margin for three examples of industrial equipment. Calculations are made for a distance of 500 m. These link margins, based on data given as an example, are basic to the understanding of laser signal attenuation by climatic phenomena (such as fog, light fog, rain, snow, and scintillation).

Emission power	6 mW (7.78 dBm)	10 mW (10 dBm)	80 mW (19.03 dBm)
Sensitivity	−38.23 dBm	−23.01 dBm	−23.01 dBm
Geometrical Attenuation (D= 500 meters)	(2nd = 1 mrad; Capture area: 0.002 m^2) 19.92 dB	(2nd = 2.5 mrad; Capture area: 0.005 m^2) 23.90 dB	2nd = 1 mrad; Capture area: 0.005 m^2) 15.94 dB
System loss	0 dB	0 dB	0 dB
Link margin	26.09 dB	9.11 dB	26.1 dB

Table 6-3: *Link margin of three typical systems for a distance of 500 m*

6.4.6.3. *Availability and quality of service*

In this section we present a concrete example of the quality of service (QoS) research for a given link using the characteristics of three examples of FSO equipment.

We consider the following elements:

− link distance: 500 meters,

− manufacturer: three manufacturers (equipment A, B and C),

− equipment: a 155 Mbps SDH interface with optical fiber,

− model: we will apply the Kruse et al. attenuation model which gives the attenuation by aerosols (fog), the most significant attenuation for a laser link,

− site: two sites are studied (Dijon and Rennes).

The research of quality of service is a process which proceeds in three steps; shown below, this process can be partially or completely computerized.

6.4.6.3.1. Example of minimum visibility calculation

We consider a 500 meter link with an interface of 155 Mbps. From technical information given by the three manufacturers relating to wavelength, emission power, sensitivity and system losses, we determine the geometrical and the molecular attenuation, then the link margin, the linear margin and, by applying the Kruse et al. attenuation model [KRUSE, 1962] [6.4], the value of the minimum visibility.

Kruse et al model

$$\beta(\lambda) \Box \; \beta_a(\lambda) = \frac{3.912}{V}\left(\frac{\lambda_{nm}}{550}\right)^{-q}$$ [6.4]

$$q = \begin{cases} 1.6 \text{ if } V > 50 \text{ km} \\ 1.3 \text{ if } 6km < V < 50 \text{ km} \\ 0.16V + 0.34 \text{ if } 1 \text{ km} < V < 6 \text{ km} \\ V - 0.5 \text{ if } 0.5 \text{ km} < V < 1 \text{ km} \end{cases}$$

	Equipment A	Equipment B	Equipment C
Wavelength (nm)	690	850	1550
Link distance (m)	500	500	500
Emission power (dBm)	10	13	26
Sensitivity (dBm)	−35	−40	−36
Geometrical Attenuation (dB)	25.94	17.4	18.59
Molecular attenuation (dB)	0.05	0.205	0.05
System losses (dB)	0	0	0
Link margin (dB)	19.01	35.395	43.36
Linear margin (dB/Km)	38.02	70.79	86.72
Value of minimum visibility (m)	341.93	183.64	149.91

Table 6-4: *Table of data*

6.4.6.3.2. Example: weather statistical data

From two weather files provides by Météo France, for Dijon and Rennes, which give the percentage fog appearance per hour, data synthesized over a long period of time; we plot graphs presenting the cumulated percentage of fog appearance in three time periods:

– 8 a.m. to 8 p.m,

– 8 p.m. to 8 a.m,

– a whole day from midnight to midnight.

These two weather files have the following characteristics:

– for Dijon:

   – hourly visibility,

   – from 100 to 8000 meters,

   – from 1992 to 2001,

   – more than 75000 observations,

   – cumulated percentage of appearance,

   – three-hourly observation periods.

– for Rennes:

   – hourly visibility,

   – from 100 to 5000 meters,

   – from 1992 to 2002,

   – more than 73000 observations,

   – cumulated percentage of appearance,

   – three-hourly observation periods.

We draw, on each graph (Figure 6.8), three curves for which we have:

– as X-coordinate: the minimum value of visibility (m) (i.e. fog density),

– as Y-coordinate: percentage of appearance of different values of minimum visibility (%).

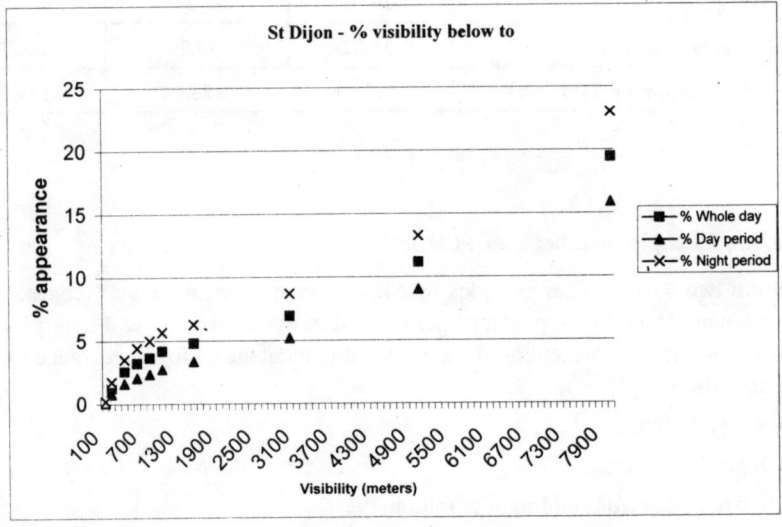

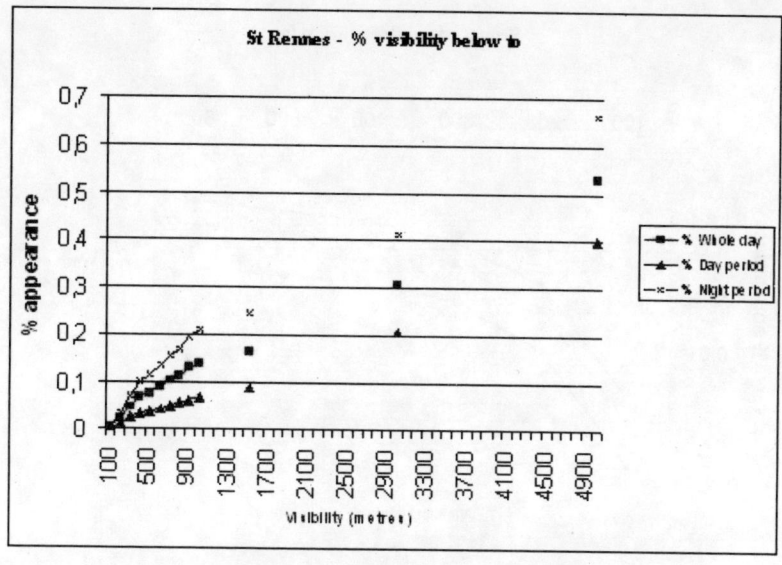

Figure 6.8: *Percentage of appearance of fog with given visibility value. Top: for the Dijon site, below: for the Rennes site*

For better legibility, the values in ordinates are presented on a logarithmic scale.

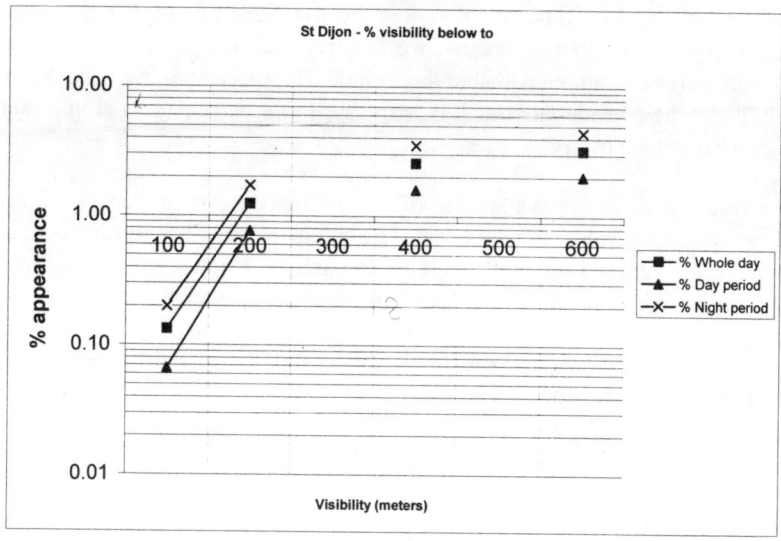

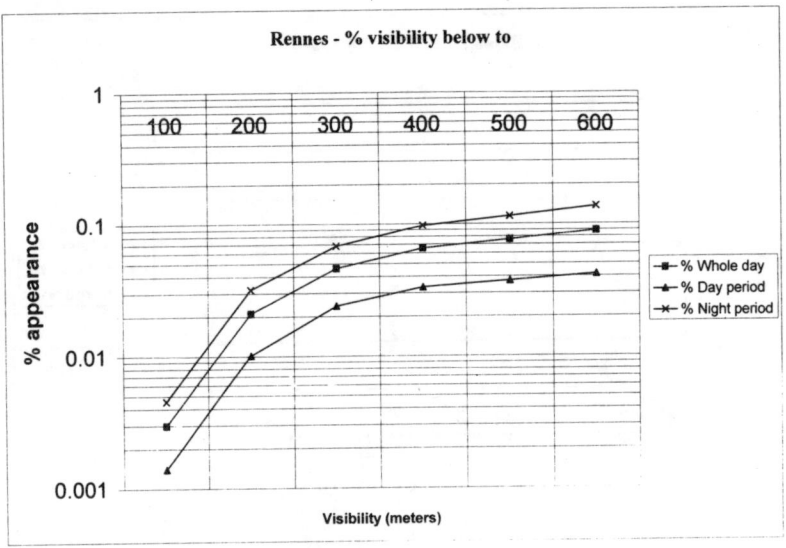

Figure 6.9: *Percentage of appearance of fog for given visibility values (logarithmic scale). Top: for the Dijon site, below: for the Rennes site*

6.4.6.3.3. Example of a link availability calculation

A laser link becomes unavailable when the fog density is higher than the link margin. Using meteorological graphs; we thus research the percentage of time for which we will have an unavailable laser link. By extension, we determine the availability of the considered laser link, expressed as a percentage and in a number of hours (or minutes) per year.

The research of the unavailability of a laser link is carried out by taking the minimum visibility value for each set of equipment (the X-coordinate) and by determining the value of the percentage of appearance. Knowing this value, we can estimate the limit beyond which the laser link becomes unavailable.

Table 6-5 below shows the percentage values and unavailability values for each set of equipment for each of the two sites.

Dijon

% appearance		Equipment A	Equipment B	Equipment C
Minimum visibility value (meters)		341.93	183.64	149.91
Model Zoom	**Day**	2.1	0.9	0.4
	P8-20	1.3	0.5	0.22
	P20-8	2.8	1.3	0.6

Rennes

% appearance		Equipment A	Equipment B	Equipment C
Minimum visibility value (meters)		341.93	183.64	149.91
Model Zoom	**Day**	0.052	0.015	0.008
	P8-20	0.028	0.007	0.004
	P20-8	0.08	0.022	0.012

Table 6-5: *Comparison of the Dijon and Rennes sites*

By extension, we can determine the availability of the considered laser link, expressed as a percentage and in a number of hours (or minutes) per year. For the two sites under consideration, see Table 6-6.

Dijon

Equipment A	**% availability**	**Unavailability - No. Hours/Year**
Year	97.9	180
8 a.m. to 8 p.m. period	98.7	57
8 p.m. to 8 a.m. period	97.2	123

8760 hours = 1 year

Equipment B	**% availability**	**Unavailability - No. Hours/Year**
Year	99.1	79
8 a.m to 8 p.m. period	99.5	22
8 p.m to 8 a.m. period	98.7	57

8760 hours = 1 year

Equipment C	**% availability**	**Unavailability - No. Hours/Year**
Year	99.6	35
8 a.m. to 8 p.m. period	99.8	10
8 p.m. to 8 a.m. period	99.4	26

Table 6-6: *Link availability on the Dijon site for 3 different sets of equipment A, B and C*

Rennes

Equipment A	% availability	Unavailability - No. Hours/Year
Year	99.948	4.56
8 a.m. to 8 p.m. period	99.972	1.23
8 p.m. to 8 a.m. period	99.920	3.50

8760 hours = 1 year

Equipment B	% availability	Unavailability - No. Hours/Year
Year	99.985	1.31
8 a.m. to 8 p.m. period	99.993	0.31
8 p.m. to 8 a.m. period	99.978	0.96

8760 hours = 1 year

Equipment C	% availability	Unavailability - No. Hours/Year
Year	99.992	0.70
8 a.m. to 8 p.m. period	99.996	0.18
8 p.m. to 8 a.m. period	99.988	0.53

Table 6-6 (continued): *Link availability on the Rennes site for 3 different sets of equipment A, B and C*

6.4.6.3.4. Example of availability according to the link distance

Another approach consists of presenting QoS, according to the link distance for the two sites.

Dijon

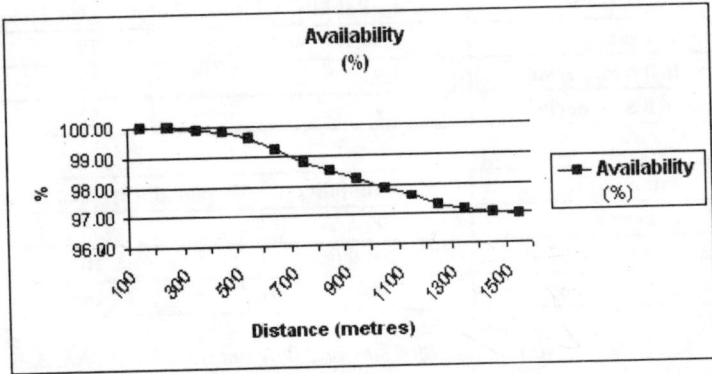

Rennes

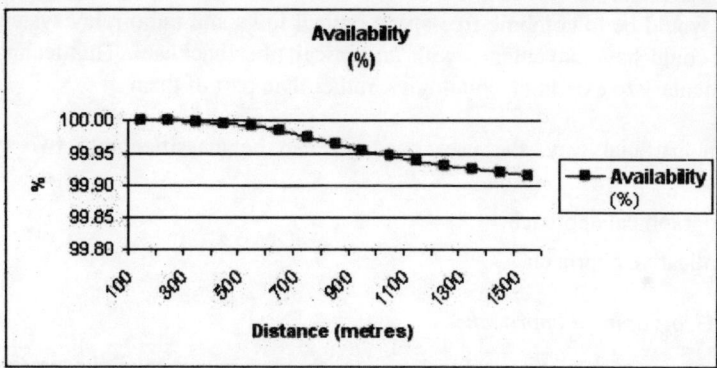

Figure 6.9: *Availability for the area of Dijon (high) and Rennes (low) according to distance*

From these two examples, we note two important factors:

− quality of service (QoS) of a laser link is highly dependent on local weather configurations (fog) and so the data-processing treatment of this climatic data becomes essential,

− quality of service management is dependent on the link distance (under identical weather conditions) and a modification or an improvement in this parameter could be achieved by the selection of equipment or manufacturer. For example, from the previous graph, if a quality of service higher than 99.95% in Rennes is required; it is necessary that the laser link does not exceed a length of 900 m.

6.4.7. *FSO potential applications*

In this section we mention the various potential applications of Free-Space Optical links. One of the principal characteristics of FSO technology is that it is transparent to the transmitted protocol, i.e. no frame is carried out on the transmission data.

The first results of the studies seem to show that this technology could answer to the famous "last mile" syndrome, i.e. the possibility to cover at high rate a restricted geographical area.

An American study results indicates that more than 95% of business zones are located at less than one mile from an optical fiber node, which indicates an important high data rate links potential for the use of point-to-point FSO link or FSO networks.

It is advisable to mention that in a network configuration example, the best solution would be to combine free-space optical links and radio-relay systems. This solution could have advantages with an optical fiber backhaul. This technology is complementary to existing technologies, rather than part of them.

In an artificial way, the uses concepts can be classified into two types of approach:

– geographical approach,

– applicative approach.

6.4.7.1. *Geographical approach*

For some specific geographical points, free-space optical links present many advantages compared to those which are proposed by radio technology or based on optical fibers:

– sites with some risk: where the geographical environment is susceptible to major weather disturbances such as lightning,

– charged electromagnetic environments: a geographical area already saturated with radio operator links, or very sensitive to radioelectric wave disturbance (for example: airports, factories),

– classified sites or private areas: a radio link solution could be impossible and the cost of optical fiber implementation prohibitive,

– not easily surmountable obstacles: over certain short distances, it is very difficult to carry out civil engineering work for laying optical fibers (example: motorways, water canals),

– dense metropolitan zones: free-space optic network installation could answer the need for a high data rate in such zones,

– hospital areas in which the use of radiocommunication is prohibited to avoid detrimental effects on people and on equipment sensitive to electromagnetic radiation.

6.4.7.2. *Applicative approach*

The second approach relates to the FSO link applicative aspect:

– factual link (example: an optical link was used between the video central control room and the Olympic swimming pool during the Olympic Games in Sydney, Australia),

– temporary link: quick and time-limited installation awaiting the installation of the final link (example: a quick commercial action to answer a customer's needs),

– emergency link: installation of an emergency link to restore a temporarily destroyed connection (example: following the September 11th 2001 attack in New

York, several optical links were established to reconnect data-processing and telephone networks),

– inter-site links, rented or private independent networks: installation of an optical link over a short distance for a data-processing connection (Ethernet: 10 Mbps, FastEthernet: 100 Mbps, GigabitEthernet: 1.0 Gbps, FFDI) or telephone (traditional or via IP protocol),

– closing an optical loop: to obtain additional safety in closing an optical fiber link (ATM: 155 Mbps, WDM: 10 Gbps),

– GSM or UMTS Inter-cell link: in urban zones, where, due to network congestion, broadcasting cells have an increasingly restricted zone (microcells or even picocells), an optical link can offer an advantageous solution, particularly for the important bandwidth requirements for the new mobile telecommunication generation (3G).

Chapter 7

Safety and Confidentiality

7.1. Safety

7.1.1. *Dangers*

Any laser can present dangers to man, both at an ocular level and a cutaneous level. In this section, we do not mention the cutaneous effects, because the energy levels of FSO equipment do not present a significant risk for the skin. However, the human eye is very sensitive to infrared radiation.

Table 7-1 shows the various effects for three ranges of wavelength.

The important factors to take into account in order to evaluate the risks are: the signal wavelength, the powers encountered, and the beam form. The characteristics of the various factors of infrared radiation have been studied in depth and there are rules linked to their use or handling. The standards in force in France are (AFNOR – CEI, 1993 and 1994):

– NF IN 60825-1: safety of laser apparatus – Part 1: materials classification, regulations and user guide (1993). Supplemented amendments A1 (1997) and A2 (2001),

– NF IN 60825-2: safety of laser apparatus – Part 2: safety of telecommunication systems using optical fibers (1994).

Wavelength	550–700 nm	700–1000 nm	1500–1800 nm
Emission	visible-red	near-infrared IR-A	far-infrared IR-B
Cutaneous effects	no significant effects		
Eye	percentage transmission and absorption of a laser signal		
Transmission	90%	50%	0%
Absorption	*retina*: 90 %	*cornea*: 0% *vitreous humor*: 50%	*cornea*: 90% *aqueous humor*: 10%
Attack of the eye (for thresholds that exceed MPE*)	*retina*: heating, burns, lesions limited but irreversible	*retina*: heating, burns, lesions limited but irreversible. *crystalline lens*: potential lesion	*Cornea*: potentially opacifying

Table 7-1: *Effects of standard semiconductor lasers on the eye*
(the definition of the MPE is given in the text)

These two standards are integral parts of the International Electrotechnical Commission (CEI 60825-1 and CEI 60825-2) publication.

It should be noted that a draft standard, which directly applies to optical links, is as follows: PR NF IN 60825-12: safety of laser apparatus – Part 12: safety of free-space optical communication systems used for the transmission of information. For more information, the AFNOR classification index is: C43-822PR.

Lasers are divided into various classes according to the risks which they are known to present. Exploiting the reference material is relatively complex because of the many parameters that have to be taken into account. The standard defines, among others, the AEL (Accessible Limits Emission) levels and the MPE (Maximum Permissible Exposure) levels.

The following parameters allow us to classify and quantify risk:

– categories, which define a risk and a power level, an obligatory description for any laser product,

– AEL, a category which corresponds to the power that is emitted by the laser apparatus,

– MPE, a risk factor, applies to the level of laser radiation to which a person may be exposed without hazardous effects or adverse biological changes in the eye or skin,

– NOHD, the Nominal Ocular Hazard Distance, is the axial beam distance from the laser where the exposure or irradiance falls below the Maximum Permissible Exposure (MPE) adapted to the level of the cornea; beyond this distance, the risk becomes negligible.

7.1.2. *Concept of categories*

The standards, in particular standard EN 60825-1, index lasers into 7 classes. Each class is defined according to power or energy values emitted by the laser and accessible by the user, they are the Accessible Emission Limits: AEL. These classes are used to determine the level of risk of laser radiation, in order to calculate or check the attenuation necessary to avoid suffering damage (attenuation due to the distance, the emitted beam width, and the angular variation).

Category	Associated risks
1M	Low power device emitting radiation at a wavelength in the band 302.5–4000 nm. Device intrinsically without danger from its technical design under all reasonably foreseeable usage conditions, including vision using optical instruments (binoculars, microscope, monocular)
1	Low power device emitting radiation at a wavelength in the band 302.5–4000 nm. Device intrinsically without danger from its technical design under all reasonably foreseeable usage conditions, with the exception of vision using optical instruments (binoculars, microscope, monocular)
2M **(IIM)**	Low power device emitting visible radiation (in the band 400–700 nm). Eye protection is normally ensured by the defense reflexes including the palpebral reflex (closing of the eyelid). The palpebral reflex provides effective protection under all reasonably foreseeable usage conditions, including vision using optical instruments (binoculars, microscope, monocular)
2 **(II)**	Low power device emitting visible radiation (in the band 400–700 nm). Eye protection is normally ensured by the defense reflexes including the palpebral reflex (closing of the eyelid). The palpebral reflex provides an effective protection under all reasonably foreseeable usage conditions, with the exception of vision using optical instruments (binoculars, microscope, monocular)
3R **(IIIR)**	Average power device emitting a radiation in the band 302.5–4000 nm. Direct vision is potentially dangerous
3B **(IIIB)**	Average power device emitting a radiation in the band 302.5–4000 nm. Direct vision of the beam is always dangerous
4 **(IV)**	High power device There is always danger to the eye and for the skin, fire risk exists

Table 7-2: *Classification of lasers, (i) there is a slight difference between the European standard and the American standard (example: Europe – class 3B/USA – class IIIB; (ii) M means "Magnifiers", i.e. "Magnifying" – the inclusion of equipment into category M is due in most cases to the width of the emitted beam, its divergence and its power; (iii) for the classes 3B (IIIB) and 4 (IV), it is recommended that there should be medical checks and specific training before installation or maintenance is carried out*

7.1.3. Accessible Emission Limits (AEL)

The AEL characterizes the power emitted by the laser equipment; it is expressed in watts. The AEL value of optical apparatus allows us to classify it. The AEL values and associated categories may be calculated for systems based on their emitted optical power and normal operating conditions.

Class	Wavelength = 850 nm	Wavelength = 1550 nm
1	$P < -6.6$ dBm $P < 0.22$ mW	$P < 10$ dBm $P < 10$ mW
2	Category reserved for the range 400–700 nm –same AEL as for class 1	
3R	-6.6 dBm $< P < 3.4$ dBm 0.22 mW $< P < 2.2$ mW	10 dBm $< P < 17$ dBm 10 mW $< P < 50$ mW
3B	3.4 dBm $< P < 27$ dBm 2.2 mW $< P < 500$ mW	17 dBm $< P < 27$ dBm 50 mW $< P < 500$ mW
4	$P > 27$ dBm $P > 500$ mW	$P > 27$ dBm $P > 500$ mW

Table 7-3: *Example of Accessible Emission Limit classification (in Watt and dBm) for two wavelengths. The power in dBm is: P (dBm) = 10 × log[P (in mW)/1 mW], i.e. +3 dBm corresponds to double the power expressed in mW.*

7.1.4. Maximum Permissible Exposures (MPE)

In order to better ensure the protection of the eye or the skin, exposure limit values were defined for people (MPE). These limit values correspond to the conditions under which the eye or the skin can be exposed without suffering lesions; specific values have been defined for the skin and the eye (reference values were selected because the thresholds are lower than those of the skin). The maximum exposure allowed is expressed in Jm^{-2}. The standards also give values of maximum illumination license which is expressed in Wm^{-2}. They are used to determine the risk level of laser radiation. These values should not be regarded as precise limit values between safety and danger levels.

Exposure duration (seconds)	1	2	4	10	100	1000	10000
MPE (Wm^{-2}) at 850 nm	36	30	25	20	11	6.5	3.6
MPE (Wm^{-2}) at 1550 nm	5600	3300	1900	1000	1000	1000	1000

Table 7-4: *Example of MPE values (in Wm^{-2}) of the eye (cornea) according to the exposure duration (in seconds) and wavelength*

When light passes from the cornea to the retina, the increase in light intensity is roughly the ratio of the surface of the pupil to the retinal image (approximately 10 µm diameter); this increase results from light focusing on a point on the retina which is entered through the pupil.

The MPE is higher for brief exposure durations than for high exposures times (for short duration exposure the tolerance will be larger); the MPE at the corneal level is much higher at 1550 nm than at 850 nm, this is related to the laser radiation absorption at the level of the various eye components; the difference in MPE values can be explained by the fact that at approximately 850 nm 50% of the signal can reach the retina whereas at 1550 nm the signal is almost completely absorbed by the cornea and aqueous humor.

7.1.5. *The NOHD calculation*

The Nominal Ocular Hazard Distance of ocular risk (NOHD) is the axial beam distance from the laser where the exposure or irradiance falls below the applicable exposure limit (MPE) adapted at the level of the cornea (see NF IN 60825-1 standard, Paragraph 3.56).

For this calculation, to neglect the atmospheric attenuation led to the consideration of the worst case. It is the adopted approach when laser safety is treated. In this case, the NOHD is represented by the following equation:

$$NOHD = \frac{\sqrt{4P_o / \pi E_{EMP}} - a}{\phi} \qquad [7.1]$$

where:

- a is the diameter of the emergent laser beam (m),
- E_{EMP} is the power density corresponding to the allowed maximum exposure (Wm^{-2}),
- P_0 is the total radiated power of the laser (W),
- ϕ is the beam divergence (rad).

Example of application for FSO products:

In Table 7-5, the a, P_0, ϕ and λ parameters appear as well as the NOHD associated to various systems. Some systems have an adjustable beam divergence. In this case, the NOHD calculation takes into account the divergence recommended by the manufacturer for the 1 km range.

Manufacturer	Product	A mm	P_0 mW	ϕ mrad	λ nm	NOHD m
A	X	150	160	8.7	1550	NV
A	Y	150	400	8.7	1550	NV
B	X	23	4.4	6	850	NV
B	Y	23	4.4	3	850	NV
C	X	21	2	3	850	NV
C	Y	21	8.8	3	850	3.7
D	X	150	100	11	910	NV

Table 7-5: *Minimal distance of ocular risk for various materials. NV: Negative value: the result of the calculation of the NOHD is negative, i.e. that equipment is a safe laser before the output of the emission lens beam*

Let us note that, as indicated in previous equations, the wavelength is significant in the NOHD calculation: the higher the wavelength, the better the laser safety criteria (all other system parameters remaining equal). If we place ourselves in the situation where the systems mentioned above all emit at 1550 nm, then these systems do not present any ocular risk whatever the considered duration of exposure.

7.1.6. *Conformity to standard IEC825/EN60825*

7.1.6.1. *Class 3B*

– Practically no products exist in 3B category. Nevertheless, for information, at the time of installation where lasers of a higher category than the 3R category are employed, a laser safety supervisor must be designated who is responsible for the following safety measures:

- use warning enclosures, if possible, and ensure the beam's path conforms to certain rules (short distances, not at normal eye level),

- affix suitable warning signs on the zone entries or the protective enclosures containing the equipment,

- limit the risk of potentially dangerous reflections and use an adequate eye guard (or protective clothing),

- ensure the person in charge of monitoring of such systems has received adequate training, and that users have appropriate medical follow-ups,

– and concerning the equipment:

- connect a remote locking system to a central emergency disconnecting locking switch or to the safety locking system of the room, door or piece of furniture,

- any apparatus must be protected from unauthorized use by the removal of the command key when it is not in service,

- avoid accidental exposure present by using a beam attenuator or by stopping the beam,

- ensure instructions accompany the equipment and contain instructions for safe use.

7.1.6.2. *Class 3R*

– Few products exist in 3R category; nevertheless, the following safety measures are essential:

- if possible, use warning enclosures and ensure the beam's path conforms to certain rules (short distances, not at normal eye level),

- the person in charge of monitoring of such systems must have received training and the users must have appropriate medical follow-up,

– And concerning the equipment:

- avoid accidentally exposing the people present by using a beam attenuator or by stopping the beam,

- ensure instructions accompany the equipment and contain instructions for safe use.

7.1.6.3. *Class 2*

– The following safety measures, concerning the use of the equipment, are prescribed:

- avoid the accidental exposure of people present by using a beam attenuator or by stopping the beam,

- instructions must accompany the equipment and contain instructions for safe use.

7.1.6.4. *Class 1*

The majority of the available products are classified as 1 or 1M (see IEC825/EN60825 standards): instructions must accompany the equipment and contain instructions for safe use.

It should be noted that the future availability of a standard, specific to free-space digital transmission optical apparatus (IN 60825-12), will allow us to better define and comprehend safety concepts in relation to the use and the environment.

7.2. Confidentiality

7.2.1. *Transmitted data confidentiality*

The majority of manufacturers use a "one-off" type amplitude modulation for data transmission by laser, the transmission protocol is generally transparent, but the possibility of "hacking" information is limited.

Apart from any direct action on the equipment or its accesses, there are only two ways for a person with important technical skills and complex interventions criteria "to recover" the transmitted data.

Information hacking is only possible under the following conditions:

1) use of the same FSO equipment, from the same manufacturer, to collect and to decode data,

2) intercept a part of the beam for data "collecting" (this is, however, very directional), and "recover" sufficient energy to process them (Figure 7.1):

– either between the two sites (A), to obtain the data transmitted from site X or Y only, with the additional difficulty which is avoiding cutting the beam,

– or, for example, behind the Y site (B) for the data transmitted from site X, knowing that the signal attenuation is very important as one moves away from the source,

– or, for example, in front of the equipment of the Y site (B) for the data transmitted from the site Y, with the additional difficulty of avoiding cutting the beam.

3) finally, the last difficulty consists of knowing the transmitted protocol, in order to interpret the collected data.

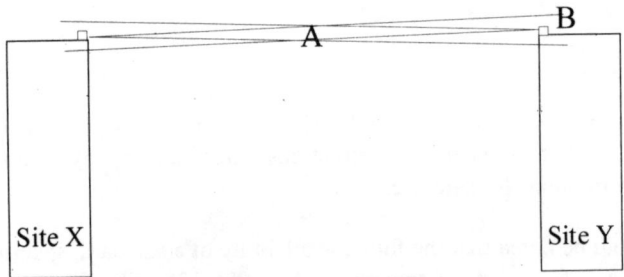

Figure 7.1: *To collect bits emitted by sites X or Y, the spy must collect part of the emission–reception beam*

7.2.2. Confidentiality techniques

The confidentiality of communication between two communicators is assured if a third person or an unspecified detector cannot collect exchanged information. To achieve this, one can either code information, i.e. use cryptography, or use material techniques to prohibit the information beam from going to specific places.

7.2.2.1. *Cryptography*

The issue of communication confidentiality can have strategic importance for economic as well as military questions; it also becomes more and more significant at the level of protection of privacy. To ensure the necessary confidentiality, the interlocutors must make the contents of their messages incomprehensible to a third person or a possible spy. The technique used is called cryptography, which literally means the art of hiding (crypto) a written text (graphic communication). Whatever the technique used in cryptography, the guiding principle is always the same: A transmitter, Alice, codes her message using a key. This results in an encrypted message, which Alice sends to her correspondent, Bob. Bob, using his key, deciphers the received message and can read it. The confidentiality of the correspondence between Alice and Bob is guaranteed owing to the fact that a third person, a spy, called for example Eve, does not have the decoding key. Admittedly, the encoding system safety should not depend on the safeguarding of the secret of the algorithm, only the secret key.

Since the beginning of cryptography, there has been an escalation of interaction between cryptographs and code breakers. When codes are no longer effective in hiding information from Eve, Alice and Bob must change their cryptographic protocol. And the protocols must become increasingly complex so that it is necessary to change them as rarely as possible.

7.2.2.2. *Public key and secret key cryptography*

In electronic communication, RSA (Rivest, Shamir, Adleman) code is currently used. This is a public key code, thus avoiding the constraint that the two interlocutors must obligatorily exchange, as a preliminary, the secret key. The safety of the algorithm rests on a mathematical conjecture, which stipulates that it is very difficult to calculate p and q prime numbers from their product. This conjecture was established from the mathematical properties of the best known factorizations algorithms and the computing power of computers. Nevertheless, it is not proven that the current factorization algorithms are optimal. According to Moore's law, computer power doubles every 18 months. Increasingly large keys are factorized. Currently, we regard those using at least 1024 bit keys as secure. However, this key could be factorized in 2048 bits, taking into account known evolutions.

Based on unproven mathematical conjectures, although simple to implement since there is no exchange of secret key, the RSA algorithm does not allow absolute communication safety. For security of exchanges, we can turn towards secret key codes. There is then no complex algorithm to use. But the key must be as long as the coded message and should be used only once. These conditions mean that this type of cryptography is used only at the diplomatic and military level.

Therefore, we have on the one hand a public key algorithm which does not require the exchange of a secret key but whose absolute safety is not guaranteed. On the other hand, we have an infallible proven algorithm, which requires the exchange of a secret key at the time of each correspondence.

7.2.2.3. *Quantum cryptography*

The only coding technique to guarantee perfect inviolability of communications is quantum cryptography. This is based on the exchange of single photons between the two correspondents. Quantum cryptography uses the Heisenberg uncertainty principle. Some quantum state characteristics cannot be measured simultaneously: whoever measures one, modifies the others. The quantum state represents the key which allows us to decode the message. In quantum key distribution, the key is sent over a quantum channel. If Eve taps the line, due to the nature of photons, she inevitably transmits errors that can be detected through the changes in the statistics of the photons Bob ultimately receives [BENNETT 1984, BENNETT 1992].

Three types of methods have been reported so far for preparing photons in the required quantum states. In the pioneering work of Bennett et al. [BENNETT 1992], the sender uses polarized photons for which the linear and circular polarization states form a pair of conjugate bases that are used to represents bits "0" and "1". The information bit is coded on the polarization state of a single photon (qubit), but this information can only be read if the polarization base in which it was coded is known.

The second method [TOWNSEND, 1993] is to use optical delays that encode each bit of information. In this case, the sender uses an interferometer.

Merolla *et al.* [MEROLLA 1999] propose a method in which Alice encodes each bit of information into sidebands of phase modulated light. The relative phase of interacting quantum states is reliably controlled by the phase of a low frequency modulating signal. This type of interference can be used to build a robust system for quantum cryptography.

7.2.2.4. *Quantum telecommunications in free space*

The property of quantum mechanics which will guarantee the safety of key transmission is the not-cloning theorem. It is not possible to copy an unknown quantum state perfectly: a spy cannot copy a qubit. One can thus distribute a coding quantum key (QKD: Quantum Key Distribution), by coding information on the polarization state of a single photon. When the key has the size of the message, the confidentiality is total; no mathematical attack can break it.

Quantum transmissions in free space have been carried out, but generally from approximations of single-photon sources. The method generally applied to simulate a single-photon source consists of strongly attenuating the emitted beam of a laser impulse. For an average number of 0.1 photon per impulse, we have 90.5% empty impulses, 9% containing 1 photon and 0.5 % containing 2 of them. These statistics of the variable number of photons corresponding to each impulse are not very satisfactory for true communication but interesting for laboratory experiments. The first experiment, in 1992, was made on a 30 cm free-space link [BENNETT, 1992]. The record known by the authors is an exchange of secret key on a 23 km free-space mountainous link [KURSIEFER, 2002]. The essential problems then were climatic fluctuations and parasitic lights.

Quantum transmissions in free-space are perfect if the luminous impulses are performed with single photons, sent one by one, of which Alice knows the polarization state. To operate these single photons, it is necessary to have sources of single photons and detectors of these single photons.

Experiments with single photons sources were performed by Beveratos et al. [BEVERATOS, 2002] and Alliaume [ALLIAUME, 2004]: NV defects in diamond emit photons with a wavelength around 637 nm; by diluting these defects in the matrix, the authors created a source of single photons. They transmitted approximately 8000 secret bits per second in a 50 meter free-space. This system allows a quantum key distribution "by request", but not at wavelengths suitable for free-space telecommunication.

Another single-photon source is proposed using erbium ions diluted in a large gap host matrix, a dielectric such as silica for example, [de FORNEL, 2001 and 2003] whose single-photon emission at 1.55 μm emitted by only one excited erbium ion is collected in near field by a thinned fiber before being re-emitted in free-space. This technique gives a genuine single-photon source at the traditional wavelength of optical telecommunications (1.55 μm) and at the ambient temperature. Thanks to the flexibility of optical fibers, we have very great simplicity of implementation which must help solve the problems of alignment between the transmitter and the detector.

7.2.2.5. Non-encrypted links in confined space: contribution of artificial materials

To guarantee the confidentiality of communication in confined space, the transfer of information must be forbidden outside the limits of the space where the communication takes place. In a conference room, researchers [de FORNEL, 2002] showed that the use of structured materials can guarantee confidentiality. The two or three dimensional structured artificial materials could become perfect reflectors. The artificial materials structured with one dimension already make it possible to have very good transmission of visible light and a very high reflection at 1.55 μm. Transmission is therefore lower than −70 dB for angles of incidence lower than 20°. But, if the angle of incidence exceeds 30°, the properties of the structure are not sufficient to secure the communications. If there is a rejecter filter at 1.55 µm whatever the angle of incidence, it is possible to use more complex structures, initially a combination of multi-layer systems of different optical and geometrical characteristics. Another alternative consists of using two or three dimensional structured materials. Photonic crystals with three dimensions make it possible to carry out perfect reflections whatever the incident angle of the light [de FORNEL, 2003].

Bibliography

[AFNOR-CEI]: NF EN 60825-1: Sécurité des appareils lasers.

[AFNOR-CEI]: NF EN 60825-2: Sécurité des systèmes de télécommunication par fibres optiques.

[ALLEAUME, 2004] R. ALLEAUME, F. TREUSSART, G. MESSIN, Y. DUMEIGE, J.-F. ROCH, A. BEVERATOS, R. BROURITUAL, J.-P. POIZAT and P. GRANGIER, "Experimental open air quantum key distribution with a single photon source", *New Journal of Physics*, 6, 92, 2004.

[ALNABOULSI, 2003a] M. ALNABOULSI, H. SIZUN, F. de FORNEL, "*Prédiction de l'atténuation d'un rayonnement laser dans le brouillard dans la région spectrale 690 à 1550 nm, Ondes Hertziennes et Diélectriques (OHD'2003)*", 3–5 September 2003, CALAIS, 2003.

[ALNABOULSI, 2003b] M. ALNABOULSI, H. SIZUN, F. de FORNEL, "*Fog attenuation of a laser radiation in the 690 to 1550 nm spectral region*", ECWT, 2003, MUNCHEN, 2003.

[ALNABOULSI, 2004] M. ALNABOULSI, H. SIZUN, F. de FORNEL, "Fog attenuation for optical and infrared waves", *Journal of Optical Engineering*, Vol. 43, 319-329, 2004.

[AUDEH, 1995] M. D. AUDEH, J. M. KAHN, "Performance evaluation of baseband OOK for wireless indoor infrared LAN's operating at 100 Mb/s", *IEEE Trans. Commun.* Vol. 43, no. 6, 2085-2094, June 1995.

[BARRY, 1991] J. R. BARRY, J. M. KAHN, E. A. LEE, D. G. MESSERSCHMITT, "High-speed nondirective optical communication for wireless networks", *IEEE Network Magazine*, 44-54, 1991.

[BATAILLE, 1992] P. BATAILLE, "*Analyse du comportement d'un système de télécommunications optique fonctionnant à 0,83 μm dans la basse atmosphere*", thesis, University of Rennes 1, 1992.

[BENNETT, 1984] C. H. BENNETT, G. BRASSARD, Proceedings of IEEE International Conference on Computers, *Systems and Signal Processing* (IEEE), NEW YORK, 1984, 175.

[BENNETT, 1992] C. H. BENNET, F. BESSETTE, G. BRASSARD, L. SALVAIL, J. SMOLIN, "Experimental quantum cryptography", *Journal of Cryptography*, 5, 3, 1992.

[BERK, 1989] A. BERK, L. S. BERNSTEIN, D. C. ROBERTSON, "*MODTRAN: a moderate resolution model for LOWTRAN 7*", GL-TR-89-0122, U.S. Air Force Geophysics Laboratory, Hanscom, AFB, Mass., 1989.

[BERTHO, 1981] C. BERTHO, "*Télégraphes et téléphones, de Valmy au microprocesseur*", Le livre de Poche, 1981.

[BEST, 1950] A. C. BEST, "Empirical formulae for the terminal velocity of water drops falling through the atmosphere", *Quart. J. R. Met. Soc.*, Vol. 76, 302-311, 1950.

[BEVERATOS, 2002] A. BEVERATOS, R. BROURI, T. GACOIN, A. VILLING, J.-P. POIZAT and P. GRANGIER, "Single photon quantum cryptography", *Physical Review Letters*, Vol. 89, 187901 (2002).

[BODEUX, 1977] A. BODEUX, "*La fréquence du brouillard en Belgique*", BRUXELLES: Institut Royal météorologique de Belgique, 1977.

[BORN, 1959] M. BORN, E. WOLF, "*Principle of optics*", Pergamon Press, London, 1959.

[BORN, 1983] M. BORN, E. WOLF, "*Principles of optics. Electromagnetic theory of propagation interference and diffraction of light*", Cambridge University Press, 6th edition, 1980.

[BRUHAT, 1992] G. BRUHAT, "*Cours de physique générale*", 6th edition, Masson, 1992.

[CARBONNEAU, 1998] T. H. CARBONNEAU, D. R. WISELY, "*Opportunities and challenges for optical wireless, the competitive advantage of free space telecommunications links in today's crowded marketplace*", SPIE conference on optical wireless communications, BOSTON, Ma, Vol. 32, 1998.

[CARRUTHERS, 1997] J. B. CARRUTHERS, J. M. KAHN, "Modeling of nondirected wireless infrared channels", *IEEE Transactions on Communications*, Vol. 45, no.10, October 1997.

[CCIR, 1990] "*Attenuation of the visible and infrared radiation*", CCIR Recommendations and reports, Vol. V, Propagation in non-ionized media, ITU, 1986.

[CHABANE, 2004] M. CHABANE, M. ALNABOULSI, H. SIZUN, O. BOUCHET, "*A new quality of service FSO software, SPIE, STRASBOURG*", France, 2004.

[CHU, 1968] T. S. CHU, D. C. HOGG, "Effects of precipitation on propagation at 0.63, 3.5 and 10.6 microns", *BSTJ*, Vol. 47, 5, 723-759, 1968.

[CEI 50, 1987] Commission électronique internationale, *Vocabulaire électrotechnique international, chapitre 845: Eclairage* CEI50, 1987.

[COJAN, 1997] Y. COJAN, J. C. FONTANELLA, "*Propagation du rayonnement dans l'atmosphère, Technique de l'ingénieur*", traité électronique, 1997.

[COZANNET, 1983] A. COZANNET, J. FLEURET, H. MAITRE, M. ROUSSEAU, "*Optique et telecommunications*", Collection Technique et scientifique des télécommunications, EYROLLES, 1983.

[De FORNEL, 1997] F. de FORNEL, "*Les ondes évanescentes en optique et en optoélectronique*", Collection Technique et Scientifique des télécommunications, EYROLLES, 1997.

[DE FORNEL, 2000] F. de FORNEL, "*Evanescent waves from newtonian optics to atomic optics*", Springer, 2000.

[DE FORNEL, 2001] F. de FORNEL, P.-N. FAVENNEC, A. RAHMANI, L. SALOMON and L. BERGUIGA, "*Source à peu de photons commandables*", French patent 00 03-96, 2000, "*Single photon source and sources with few photons*", US patent WO 01/69841 A1, 2001.

[DE FORNEL, 2002 and 2003] F. de FORNEL, R. MOUSSA, L. SALOMON, C. BOISROBERT, H. SIZUN and P. GUIGNARD, "*Matériaux artificiels pour les communications sécurisées*" in *objets communicants*, edited by C. Kintzig *et al.*, Collection Technique et Scientifique des Télécommunications, Hermès Lavoisier, 2002.

[DE FORNEL, 2003] F. de FORNEL, R. MOUSSA, L. SALOMON, C. BOISROBERT, H. SIZUN and P. GUIGNARD, "*Artificial materials for protected communications in Communicating with smart objects*", edited by Kintzig *et al.*, Kogan Page Science (London), 2003.

[DE NANSOUTY, 1911] M. de Nansouty, "*Electricity*", 1911.

[DION, 1997] D. Dion, "Synthèse sur la propagation optique en atmosphère marine", in *Propagation électromagnétique du décamétrique à l'angström*, RENNES, October 1997.

[FAVENNEC, 1996] P.-N. FAVENNEC, "*Technologies pour les composants à semiconducteurs – principes physiques*", Masson, 1996.

[FREEMAN, 1960] M. H. FREEMAN, "*Optics*", Butterworths ed, 1990.

[FUJIWARA, 1960] M. FUJIWARA, "An analytical investigation of the variability of size distribution of raindrops in convective storms", Proc., *Weather Radar 8th Conf.*, 159-166, 1960.

[GATHMAN, 1983] S. G. GATHMAN, "Optical properties of the marine aerosols as predicted by the Navy aerosol model", *Opt. Eng.*, (22), 57-62, 1983.

[GATHMAN, 1989] S. G. GATHMAN, "*A preliminary description of NOVAM, the Navy Oceanic Vertical Aerosol Mode*", NRL R-9200, 1989.

[GFELLER, 1979] F. R. GFELLER, U. BAPST, "Wireless in-house data communication via diffuse infrared radiation", *Proceedings of the IEEE*, Vol. 67, no. 11, 1979.

[GUERRIE, 1951] M. GUERRIE, "*Les ondes et les hommes*", Julliard, 1951.

[HASHEMI, 1994] H. HASHEMI, G. YUN, M. KAVEHRAD, F. BEHBAHANI, P. A. GALKO, "Indoor propagation measurements at infrared frequencies for wireless local area networks applications", *IEE Transactions on Vehicular Technology*, Vol. 43, no. 3, 1994.

[HUDSON, 1969] R. D. HUDSON, *"Atmospheric transmittance measured over 1820m path at sea level"*, Jr. Infrared System Engineering, Wiley & Sons, 115, 1969.

[JOINDOT, 1996] I. and M. JOINDOT (and 12 co-authors), *"Les télécommunications par fibres optiques"*, Collection Technique et Scientifique des Télécommunications, Dunod, 1996.

[JOSS, 1968] J. JOSS, J. C. THAMS, A. WALDVOGEL, *"The variation of raindrop size distribution at Locarno"*, Proc. of international conference on cloud physics, Toronto (Canada), 1968.

[KAHN, 1995] J. M. KAHN, W. J. KRAUSE, J. B. CARRUTHERS, "Experimental characterisation of the non directed indoor infrared channels", *IEEE Transactions communications*, Vol. 43, nos 2/3/4, 1995.

[KAHN, 1997] J. M. KAHN, J. R. BARRY, "Wireless Infrared Communications", *Proceedings of the IEEE*, Vol. 85, no. 2, 1997.

[KHRGIAN, 1952] A. K. A. KHRGIAN, I. P. MAZIN, "The drop size distribution in clouds", *Tr. Tsentro. Aerol. Obs.*, 7, 56-61, 1952.

[KIM, 2001] I. I. KIM, B. McARTHUR, E, J, KOREVAAR, "Comparison of laser beam propagation at 785 nm and 1550 nm in fog and haze for optical wireless communications", *Proc. SPIE*, 4214, 26-37, 2001.

[KNEIZYS, 1983] F.X. KNEIZYS *et al.*, *"Atmospheric transmittance/radiance: computer code MODTRAN 6 AFGL-TR 83-0887"*, Air Force Geophysical Laboratory, LEXINGTON MA, USA, 01731, 1983.

[KRUSE, 1962] P. W. KRUSE, L. D. McGLAUCHLIN, R. B. McQUISTAN, *"Elements of infrared technology: Generation, transmission and detection"*, J. Wiley and Sons, NEW YORK, 1962.

[KURTSIEFER, 2002] C. KURTSIEFER, P. ZARDA, M. HALDER, H. WEINFURETER, P. M. GORMAN, P. R. TAPSTER, R. RARITY, "A step towards global key distribution", *Nature 41*, 9, 450, 2002.

[LAVERGNAT, 1997] J. LAVERGNAT, M. SYLVAIN, *"Propagation des ondes radioélectriques, Introduction"*, Collection Pédagogique de Télécommunication, Masson, 1997.

[LAW, 1943] J. O. LAWS, D. A. PARSON, "The relation of raindrop size distribution to intensity", *Trans. Amer. Geophys. Union*, Vol. 24, 452-460, 1943.

[LEE, 1994] E. A. LEE, D. G. MESSERSCHMITT, *"Digital Communications"*, 2nd edition, Kluwer, Boston, 1994.

[LEPROUX, 2001] P. LEPROUX, *Conception et optimisation d'amplificateurs optiques de puissance à fibres double gaine dopées Erbium,* thesis, University of Limoges, 2001.

[LIBOIS, 1994] L.-J. LIBOIS, "Les télécommunications, technologies, réseaux, services", *Collection technique et scientifique des télécommunications*, Masson, 1994.

[LOMBA, 1998] C. R. LOMBA, R. T. VALADAS, A.M. de OLIVEIRA DUARTE, "Experimental characterisation and modelling of the reflection of infrared signals on indoor surfaces", *IEE Proc. Optoelectron.*, Vol. 145, no. 3, June 1998.

[LOW, 1992] T. B. LOW, N. G. LOEB, "A specific marine boundary layer aerosol model", *SPIE* CP-1688, 99-109, 1992.

[MARRAKECH] Plenipotentiary Conference of the International Telecommunication Union (2002) – *Resolution 118 – Use of spectrum at frequencies above 3000 GHz.*

[MARSH, 1994] G. W. MARSH, J. M. KAHN, "50 Mb/s diffuse infrared free space link using On-Off Keying with decision feedback equalization", *IEEE Photonics Tech. Letters*, Vol. 6, no. 10, 1268-1270, 1994.

[MARSHALL, 1948] J. S. MARSHALL, W. McK PALMER, "The distribution of raindrops with size", *J. Meteorol.*, Vol. 5, 165-166, 1948.

[McCULLAGH, 1994] M. J. McCULLAGH, D. R. WISELEY, "155 Mb/s optical wireless link using a bootstrapped silicon APD receiver", *Electron. Lett.*, Vol. 30, no. 5, 430-432, 1994.

[MEROLLA, 1999] J. M. MEROLLA, Y. MAZURENKO, J. P. GOEDGEBUER, L. DURAFFOURG, H. PORTE, W. RHODES, *Physical Review*, 60, 3, 1999.

[MIDDLETON, 1952] W. F. K. MIDDLETON, "*Vision through the atmosphere*", University of Toronto Press, Toronto, 1952.

[MIE, 1908] MIE, *Ann. Phys.*, 25, 377-445, 1908.

[MOREIRA, 1997] A. J. C. MOREIRA, R. T. VALADAS, A. M. De OLIVEIRA DUARTE (1997.), "Optical interference produced by artificial light", *Wireless Networks*, 3, Issue 2, 131-140, Kluwer Academic Publishers, Hingham, USA, 1997.

[NICHOLLS, 1996] P. J. NICHOLLS, S. D. GREAVES, R. T. UNWIN, *IEE Colloquium Optical Free Space Communication Links*, Vol. 4, no. 1, 1996.

[NICODEMUS, 1977] F. E. NICODEMUS, J. C. RICHMOND, J. J. HSIA, I. W. GINSBERG, T. LIMPERIS, "Geometrical considerations and nomenclature for reflectance", *Tech. Rep. NBS Monograph 160*, National of Bureau of Standards, US Department of Commerce, Washington, DC, October 1977.

[O'BRIEN, 1970] H. W. O'BRIEN, "Visibility and light attenuation in falling snow", *J. Applied Meteor*, Vol. 9, 671-683, 1970.

[OLSEN, 1978] R. L. OLSEN, D. V. RODGERS, D. B. HODGE, "The aR^b relation in the calculation of rain attenuation", *IEEE trans. Antennas propag.*, AP-26, 318-329, 1978.

[OMM, 1989] Organisation météorologique mondiale: *Guide du système mondial d'observation*, OMM no. 448, Geneva, 1989.

[OMM, 1990a] Organisation météorologique mondiale, *Guide des systèmes d'observation et de diffusion de l'information météorologique aux aérodromes*, OMM 731, Geneva, 1990.

[ONTAR, 1999] Fascode Atmospheric Code: PcLnWin/fascode 3P, edited by ONTAR Corporation, 9 Village Way, North Andover, Massachussetts, USA, 1999.

[PEREZ-JIMENEZ, 1995] R. PEREZ-JIMENEZ, V. M. MELIAN, M.J. BETANCOR, *Analysis of multipath response of diffuse and quasi-diffuse optical links for IR-WLAN*, Proc. INFOCOM'95, Boston, MA, USA, 7d.3.1-7d.3.7, 1995.

[PEREZ-JIMENEZ, 1997] R. PEREZ-JIMENEZ, J. BERGES, M. J. BETANCOR, "Statistical model for the impulse response on infrared indoor channels", *Electronics Letters*, Vol. 33, no.15, 1997.

[PHONG, 1975] B. T. PHONG, "Illumination for computer generated pictures", *Comm. ACM*, Vol. 18, no. 6, 311-317, 1975.

[PRUNNOT, 2003] J.-C. PRUNNOT, A. MIHAESCU, C. BOISROBERT, G. NORMAND, P BESNARD, P. PELLAT-FINET, P. GUIGNARD, F de FORNEL, F. BOURGARD, "INDEED: high bit rate infrared communications in the indoor context", in *Communicating with smart objects*, edited by Kintzig C. *et al.*, Kogan Page Science (London), 2003.

[RAMINEZ-INIGUEZ, 1999] R. RAMINEZ-INIGUEZ, R. J. GREEN, *Indoor optical wireless communications*, IEE, Savoy Place, LONDON WC2R 0BL, UK, 1999.

[ROORYCK, 1986] M. ROORYCK, M. JUY, "*Prévision de l'affaiblissement dû à la pluie sur les faisceaux hertziens de la France continentale*", NT/PAB/ETR/701, 1986.

[ROTHMAN, 1987] L. S. ROTHMAN *et al.*, "*The HITRAN database: 1986 edition*", Appl. Opt., Vol. 26, no. 19, 1987.

[SEELY, 1979] S. SEELY, A. POULARIKAS, *Electromagnetics*, Marcel Dekker ed., ISBN: 0-8247-6820-5, 1979.

[SHEPPARD, 1983] B. E. SHEPPARD, "*Adaptation to MOR, Preprints of the fifth symposium on meteorological observations and instrumentation*", 11-15, 226-269, Toronto, April 1983.

[SMYTH, 1993] P. P. SMYTH, M. McCULLAGH, D. WISELEY, D. WOOD, S. RITCHIE, P. EARDLEY, S. CASSIDY, "Optical wireless local area networks – enabling technologies", *BT Technol.*, J., Vol. 11, no. 2, 56-64, April 1993.

[SMYTH, 1995] P. P. SMYTH, P. L. EARDLEY, K. T. DALTON, D. R. WISELEY, P. McKEE, D. WOOD, "Optical wireless: a prognosis", *SPIE Proc. on Wireless Data Transmission*, Vol. 2601, Philadelphia, PA, 212-215, October 1995.

[STREET, 1997] A. M. STREET, P. N. STAVRINOU, D. C. O'BRIEN, D. J. EDWARDS, "Indoor optical wireless systems – a review", *Optical and Quantum electronics* 29. 349-378, 1997.

[SURREL, 2000] J. SURREL, "Radiométrie-Photométrie", extracted from *Fondamentaux de l'Optique*, Optique et Photonique, 1, 2000.

[SZAJOWSKI, 1998] P. F. SZAJOWSKI, G. NYKOLAK, J. J. AUBORN, H. M. PRESBY, G. E. TOURGEE, E. KOREVAAR, J. SCHUSTER, I. I. KIM, "2.4 km free-space optical communication 1550 nm transmission link operating at 2.5 Gbits/s – Experimental results" *SPIE conference on optical wireless communications*, Boston, Massachusetts, Vol. 3532, 1998.

[TOFFANO, 2001] Z. TOFFANO, *Optoélectronique, composants photoniques et fibres optiques, les cours de l'Ecole Supérieure d'Electricité*, ELLIPSE, 2001.

[TOWNSEND, 1993] P. D. TOWNSEND, J. G. RARITY, P. R. TAPSTER, *Optics Letters*, 20, 1695, 1993.

[TRAVIS, 1996] TRAVIS B., EDN 41, vol. 63, 1996.

[UIT-R P.840] *"Attenuation due to clouds and fog"*, *Rec. ITU-R* P. 839, 1999.

[VASSALO, 1980] C. VASSALO, "Electromagnétisme classique dans la matière", *Collection Technique et Scientifique des télécommunications*, Dunod, 1980.

[VASSEUR, 1997] H. VASSEUR, C. OESTGES, A. VANDER VORST, "Influence de la troposphère sur les liaisons sans fil aux ondes millimétriques et optiques", in *Propagation électromagnétique du décamétrique à l'angström*, Rennes, 1997.

[VEYRUNES, 2000] O. VEYRUNES, *Influence des hydrométéores sur la propagation des ondes électromagnétiques dans la bande 30-100 GHz: études théoriques et statistiques*, thesis, University of Toulon and of Var, 2000.

[WEICHEL, 1989] H. WEICHEL, *"Laser beam propagation in the atmosphere"*, Roy F. Potter, Series Editor, The International society for Optical Engineering, Bellingham, Washington, USA, 1989.

[WONG, 2000] K. K. WONG, T. O'FARRELL, M. KIATWEERASAKUL, "Infrared wireless communication using spread spectrum techniques", *IEE Proc. Optoelectron.*, Vol. 147, no.4, August 2000.

[YANG, 2000] H. YANG, C. LU, "Infrared wireless LAN using multiple optical sources", IEE Proc. *Optoelectron.* Vol. 147, no.4, August 2000.

Index